Book 2

Sciencewise

Discovering Scientific Process through
Problem Solving

Series Titles:
Sciencewise Book 1 • Sciencewise Book 2 • Sciencewise Book 3

written by
Dennis Holley

illustrated by
Kate Simon Huntley

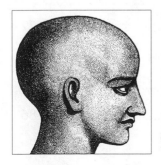
© 2002, 1996
CRITICAL THINKING BOOKS & SOFTWARE
www.criticalthinking.com
P.O. Box 448 • Pacific Grove • CA 93950-0448
Phone 800-458-4849 • FAX 831-393-3277
ISBN 0-89455-648-7
Printed in the United States of America

Table of Contents

Creative Challenges

This book is dedicated to
my wife, for her patience and understanding;
my parents, for encouraging me to wonder and explore;
my students, for teaching me more than they will ever know.

Introduction

The most beautiful thing we can experience is the mysterious. It is the source of all true art and science.

—Albert Einstein

There seem to be two aspects to the uniquely human enterprise we call science. One aspect is the continual search for natural truths. The other aspect is the vast matrix of facts and knowledge. Unfortunately, many parents and students, most textbook publishers, and even some teachers have the mistaken notion that science is memorizing terms, names, and facts. Nothing could be further from the truth.

Science is built up with facts, as a house is with stones. But a collection of facts is no more a science than a heap of stones is a house.

—Jules Henri Poincare

Science cannot be gleaned from the glossy pages of a pricey textbook. Science must be experienced, not memorized. Facts and knowledge are really the accumulated fruits of centuries of scientific labor. Make no mistake, humans are able to use these facts and this knowledge in the most incredible and creative ways imaginable, but facts alone are not science.

Imagination is more important than knowledge.

—Albert Einstein

The true essence of science is the relentless and unwavering need to know "why." It is this need, this nagging, irresistible curiosity to search for answers, that drives our species from the depths of the ocean to the blackness of outer space. This is what science really is, and kids of any gender, color, and practically any age, with the proper guidance, can <u>do</u> real science.

Unfortunately, not only do students come to you not fully understanding science, they also lack the skills necessary to <u>do</u> science. They do, however, come through your classroom door fully charged with the most important precursor to scientific inquiry—curiosity. With this remarkable inquisitiveness, students, with your help, can begin to learn the skills necessary to <u>do</u> science. There are five basic science skills students need to develop: observing, predicting, designing/experimenting, eliminating, and drawing conclusions.

Observing

Science done right requires students to be accurate and thorough observers. Learning good observation techniques requires much practice. Hand in hand with the development of these skills must come the realization that any observation is the unique perspective of the observer, colored and altered by his/her experiences, expectations, and emotions. Students must constantly be challenged to separate inference (what they think is there) from reality (what is actually there) in their observations.

In the field of observation, chance favors the prepared mind.

—Louis Pasteur

Predicting

Curiosity raises questions. Careful observation reveals information. Using this information, we make predictions about the possible answers to our questions.

Prediction is probably the easiest skill

for students to master. The main problem you will encounter is that students are often reluctant to make predictions for fear of being wrong. This is an attitude you must constantly strive to change. In science, right or wrong predictions don't matter. Science is the search for natural truths, and it matters not whether you come in the front door (correct predictions) or the back door (incorrect predictions) of the house of truth. What matters is that either way, in the end, you learn the truth.

Along with the development of this skill and this attitude must come the realization that certain predictions are more valid than others. A hypothesis is merely a guess, conjecture, or untested speculation. A theory is a higher level of prediction because it is an educated guess based on some evidence or past experience.

Designing/Experimenting

With the realization that speculation and prediction must be tested in controlled experiments to determine the truth, modern science was born. For too long, people accepted the musings of authority figures as truth and fact. Often the more bizarre the speculation, the more eager people were (and some still are) to believe it. Experimentation is what separates science from philosophy and superstition.

The practice of science enables scientists as ordinary people to go about doing generally ordinary things that, when assembled, reveal the extraordinary intricacies and awesome beauties of nature.
—Arthur Kornberg

The first hurdle to clear in experimental design is to determine what problem is to be solved. Problems should always be stated in question form, be as simple and specific as possible, and address only one factor at a time.

Let us use a whimsical imaginary problem to demonstrate experimental design. Suppose you had to solve the following problem: What is the effect on aardvarks of eating chocolate pudding? Assuming you had unlimited resources, your experimental design might go something like this:

Step 1
Determine what problem is to be solved. The problem—What is the effect on aardvarks of eating chocolate pudding?—is in question form, is specific, and deals with only one problem, so we are ready to proceed.

Step 2
Get 1,000 aardvarks. The more experimental subjects you work with, the more reliable is your data.

Step 3
Separate the aardvarks into two equal groups. In each group, put 500 aardvarks with the same age, sex, and physical characteristics. You want the two groups to be as nearly equal in all respects as possible.

Step 4
Keep both groups under identical conditions—same size cages, same amount and kind of food, same amount of water, same period of light and dark, same temperature, same humidity, and so on. These conditions, called the control variables, must be kept as nearly identical as possible.

Step 5
Feed one group of aardvarks chocolate pudding, and designate it the experimental group. Do not feed chocolate pudding to the other group, and designate it the control group. The chocolate

pudding in the experimental group is called the *manipulated variable*; it is the effect of this pudding that you are testing.

Step 6
Let the experiment run for a reasonable length of time. Collect appropriate data; for example, after two weeks the experimental group turns green and starts to do back flips.

Step 7
To verify these results, repeat the experiment as many times as possible with different aardvarks and different batches of pudding.

Step 8
Based on the accumulated data, draw reasonable conclusions. The reasonable conclusion here would be that apparently chocolate pudding causes aardvarks to turn green and do back flips.

Some problems are difficult or impossible to test experimentally. For example, to calculate the orbit of a comet, we would deduce certain outcomes then look to nature for verification.

> *The only solid piece of scientific truth about which I feel totally confident is that we are profoundly ignorant about nature.*
> —Lewis Thomas

Eliminating
Not only will you have to battle students' fear of making wrong predictions, you will also have to deal with students' fear of failure. Students must come to understand and believe that failure in science is not to be feared. Actually, we learn more from failure than success because failure raises more questions than success, and these questions, in turn, force even more inquiry. Science is dynamic and always changing. Newly discovered "facts" and even those laws of science that have withstood the test of time must be subject to revision at any moment. What we regard as facts are at best momentary illusions seen through a veil of ignorance. Today we laugh at the idea that earlier people thought it factual that the earth was flat or that living things could arise spontaneously from dead or inorganic matter. The future will show many of our so-called facts to be just as wrong.

> *It is possible that every law of nature so far has been incorrectly stated.*
> —J.B.S. Haldane

Students need to learn what data is appropriate to collect and how to organize data. Charting and graphing skills are essential. Organizing data into tables, charts, and graphs enables us to view the results in a graphic format. In this form, data is easier to understand and patterns are more easily discerned. Students must learn to deal with the fact that data may be "muddy"—inconsistent, unexpected, and often unfathomable.

Drawing Conclusions
What does the data mean? This is often difficult for professional scientists to answer, let alone students. "Muddy" data can only

yield "cloudy" conclusions, and students must learn to deal with this frustrating problem. Only experience will allow students to identify and support those conclusions that are valid and discard those that are not valid.

> *The art of becoming wise is the art of knowing what to overlook.*
> —William James

Once students have some proficiency in the above skills, they can begin to think scientifically and actually <u>do</u> science using the method illustrated below:

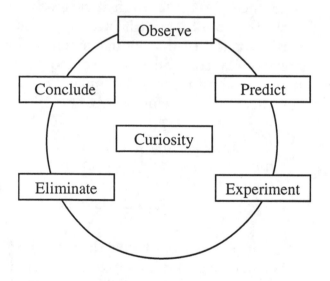

The scientific process is cyclic and becomes progressively more refined in view of the problem you are trying to solve.

Scientists approach problems in an organized way, but hunches and serendipity also play a role. What do Velcro, penicillin, x-rays, and Teflon have in common? Serendipity!— the lucky accident of finding valuable things not specifically sought after.

Science can be tedious, frustrating, and just plain boring with repeated failure often the norm. It can also be the most challenging, stimulating, and rewarding thing a person

can do. The only way for students to learn what science is and how it works is to let students actually <u>do</u> science.

> *Tell me, I forget*
> *Show me, I remember*
> *Involve me, I understand*
> —Chinese proverb

Dynamo Demos

The teacher-led Dynamo Demos (pp. 1–123) help students to develop the science process skills: observing, predicting, experimenting, eliminating data, and drawing conclusions. In addition, students develop their problem solving and creative/critical thinking skills.

In the Dynamo Demo activities, students do the thinking and the teacher does the doing. The teacher sets up and presents a "What will happen if...?" or "Why did that happen?"problem situation. Using guided questions and the necessary manipulation of apparatus and equipment, the teacher helps the students understand the problem, make accurate observations and reasonable predictions, and arrive at a conclusion or an answer to the problem.

While specific scientific principles and concepts are demonstrated in the Dynamo Demos activities, the primary focus is on actively involving students in the scientific process and developing problem solving and creative/critical thinking skills.

When possible, the Dynamo Demos are counterintuitive and present discrepant results. This approach makes the Demos more challenging and often results in incorrect predictions, forcing students to "come in the back door" to the truth.

The first demonstration is a preliminary activity designed to refine students' observation skills. The rest of the demonstrations can be done in any order.

> We must not confuse the results of science with the ways in which scientists produce those results. The practice of science is a social process, an art or a craft and not a science.
> —Michael Klapper

Creative Challenges

The student-centered Creative Challenges (pp. 124–64) help students develop their creative/critical-thinking, problem-solving, and "inventioneering" skills.

In the Dynamo Demos, the teacher sets up an experiment and uses select questions to guide the students to a solution of the given problem. With the Creative Challenges, the teacher presents the problem (challenge) then functions merely as a technical advisor. Stu-

dents must design and develop a solution to the problem (challenge). This allows students to practice and apply the science process skills demonstrated in the Dynamo Demos.

The student combines a process of "solution generation" with trial and error to look for possible solutions to a given problem. Each time a possible solution is generated, the student must predict what will happen and then observe the results when the solution is implemented. Each solution that is tried requires analysis and interpretation of the results and allows the student to draw conclusions on how to best solve the original problem. Specific scientific principles and concepts are learned through the process itself rather than through direct teacher instruction.

> Invention is a combination of brains and material. The more brains you use, the less material you need.
> —Charles F. Kettering

On Using This Book

The following are general information, guidelines, and hints for using this book:

- Use this book in conjunction with any commercial or teacher-authored science curriculum.

- The activities are designed to come from "left field." That is, the activities are not intended to teach any concept being discussed in class. The less the Demos and Challenges have to do with what is going on in class, the better. Students are then forced to think and create rather than turn to a book for the answer.

- The activities are not written to be formally graded. Assess and/or grade them as you see fit. I, personally, do not grade my students on the Demos or Challenges as I am not sure how you accurately evaluate and assess creativity and critical thinking. Furthermore, I worry that formal grading will stifle creativity and make these activities just another educational burden for students to endure.

- These activities produce creative effort and critical thinking, not marketable results. Failure and incorrect predictions are expected and even encouraged. Such is the reality of actual everyday science.

- Students should work individually as much as possible. If time and/or materials present a problem, students may work in design teams of two, preferably, but three if necessary. The larger the design teams, the less each team member will get from the activity and the more conflicts you will have within groups.

- Many of the challenges have a competitive element. Students need to be competitive and learn how to deal with competition; however, students should not lose sight of the fact that competition is secondary in these activities. Creative thinking and effort are the primary goals. Stress the process, not the product.

- Encourage students to do their own thinking and creating. Don't let competitive fervor lead to outside input by parents, siblings, and/or peers. Along those same lines, set spending limits on those challenges where students are required to supply their own construction materials.

- Try the activities before you present them to your class. The activities in this book are tried and tested, and I have attempted to write the directions in clear and concise terms. But play it safe, work things out beforehand. Don't waste precious class time by "winging it."

- Once Demos and Challenges are complete, encourage students, when practical, to take this learning out of the classroom and share it with parents, siblings, and friends. This is what I call the "Ripple Effect." Educational research shows that we best learn what we teach to others. Always consider the safety, practicality, and the necessity of supervision when having students do these activities outside the classroom.

- Give students recognition for their efforts. This might include displaying their creations and inventions, awarding and displaying appropriate certificates of achievement, and/or presenting gag gifts as trophies. Such recognition will generate a great deal of student enthusiasm, parental support, and positive public relations.

- Kids may clamor to do these activities every day, but that is not the intent of this book. This book is designed to supplement an established curriculum. You should do one or two Demos first before starting students on the individual Challenges. There are enough Dynamo Demos in this book to do one demo every week and enough Creative Challenges to do one every other week during a normal-length school year. You certainly can do fewer than this, but in my experience, it may not be practical to do more from a time standpoint.

- Ask, don't tell. It is crucial that you help students develop their thinking skills by asking for student observations, predictions, and explanations rather than just giving the answer. Mix closed questions that demand factual recall with open questions that are divergent and thought-provoking. Use questions to generate discussion, but don't always call on the student who raises his hand, and give adequate wait time when you question. Space does not permit a detailed discussion here of how to construct good oral questions, but there are excellent resources available to help you develop sound questioning strategies.

In closing, be creative yourself with this book. Take my offerings and put your own twist to them. If you come up with a better way of doing some of the things in this book or with whole new activities that could be included in future editions of this book, contact me.

Dennis Holley
Critical Thinking Books & Software
P.O. Box 448
Pacific Grove, California 93950

Dynamo Demos

1 The Mysterious Can

Problem: *A can that rolls backwards and even uphill? How can this be?*

Observe

1. In the space below, diagram and describe the system you see before you.

Predict

2. Predict what will happen when the teacher rolls the can.

Conclude

3. What did the can do when the teacher rolled it?

Predict

4. Predict what will happen when the teacher releases the can at the bottom of the ramp.

Conclude

5. Did the can behave the same as on the floor (or tabletop) or differently when the teacher released it at the bottom of the ramp? What is your explanation for what the can did?

1 For the Teacher

Objective

In this activity, students, without being able to see inside the can, will determine how a can is able to roll backwards and uphill.

Materials Needed

- 1 coffee can with opaque plastic lid
- 1 large rubber band
- 1 small fishing weight or rock
- a short piece of string
- 2 paper clips
- scissors and metal punch
- books and a board, or other materials suitable for constructing a ramp

Curiosity Hook

Have the come-back can system assembled and sitting where students can see it as they enter the classroom.

Setup

1. Punch a small hole in the center of the bottom of the can. Punch another small hole in the center of the plastic lid.

2. Fasten the weight or rock to the center of the rubber band with a piece of string. The weight or rock should hang down slightly from the rubber band.

3. Slip one end of the rubber band through the hole in the bottom of the can. Secure the loop with a paper clip to keep it from slipping back into the can. Now, slip the other end of the rubber band through the hole in the lid of the can; secure this loop with a paper clip.

4. Place the lid on the can. The rubber band must be stretched enough to hold the lid on securely. The completed system should look like this:

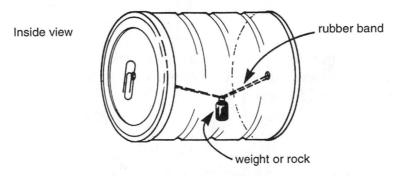

Inside view

rubber band

weight or rock

5. Have students carefully observe the come-back can system and then accurately describe it under 1 on the student pages.

6. Have students predict under 2 on the student pages what will happen when you roll the can. Now, roll the can on the floor or a tabletop. The can should roll a short distance, stop, and then roll back to you. Have students describe the behavior of the can under 3 on the student pages.

7. Construct a ramp on a tabletop. Do not make the ramp too steep. This time ask students to predict what will happen if you release the can at the bottom of the ramp. Have students write their predictions under 4 on the student pages. While they are writing, unobstrusively wind up the rubber band by turning the plastic lid; you may need to step out of the classroom to do so.

8. Place the come-back can system at the bottom of the ramp. The can should roll uphill. Have students describe the behavior of the can under 5 on the student pages.

Safety Concerns

Be careful when stretching the rubber band.

Outcomes and Explanations

1. Rolling the can along the floor causes the rubber band to twist in a forward direction. The increasing tension in the twisting rubber band applies force to the can, causing it to stop. As the rubber band starts to unwind, the tension is released in the opposite direction, and the can begins to roll backwards. The same principle applies when you turn the can's lid before placing the can on the ramp.

2. As part of the explanation, show students the inside of the can or a diagram of the inside of the can. Have students write the explanation for the can's behavior under 5 on the student pages.

Application

Challenge students to investigate the following problems:

1. Does the come-back can system operate on a one-to-one basis? That is, if you roll the can forward one foot will it roll backward one foot?

2. How much power does the system generate? Students could test this by seeing how steep a ramp the can system will climb backwards.

Take Home

If students have the materials to construct a come-back can system at home, encourage them to do so. Such a system would be an excellent way for them to amaze and teach their parents, siblings, and friends.

2 Invisible Crushing Fingers

Problem: *Can something invisible crush a can?*

Observe

1. In the space below, observe and describe what happens to the can.

Conclude

2. Why did this happen? What are the invisible "fingers" that crushed the can?

3. Suppose you were in a hurry; how could you get this process to happen faster?

2 For the Teacher

Objective

In this activity, students will use their powers of observation and critical thinking to understand how the pressure of seemingly invisible, weightless air can crush a can.

Materials Needed

- 2 empty metal cans that can be sealed airtight and heated
  (Any size can will work, but a gallon [4–5 liter] can is ideal. Smaller cans are hard for students to observe, and larger cans may be difficult to manipulate and seal. While size is not crucial, it is imperative that the cans be sealed airtight for this demonstration to succeed. Forcefully blowing into the can while sealing the opening with your lips will pressurize the can and reveal any leaks. Another method to find leaks would be to fill the cans completely full of water and seal them. Carefully dry off the outside of the cans and set them on their bottoms on a dry piece of paper. After a time, invert the cans and prop them up so that the tops become the bottoms. If they hold water, they will probably hold air.)

- a heat source—hot plate, bunsen burner, or stove

- 1 solid rubber stopper of the proper size to seal the cans

- a towel or oven mitts to prevent burning your hands as you seal the cans

- 1 aluminum pie pan or some other heat-resistant pad to set the hot cans on

- a sink or dishpan

- water to pour over one can

- safety goggles

Curiosity Hook

Start to "cook" one can on the heat source as students enter the classroom.

Setup

1. Put enough water in the can to just cover the bottom.

2. Place the unsealed can on the heat source. Heat until the water in the can is boiling vigorously. You should be able to hear the water bubbling in the can, and steam should be flowing out the opening of the can.

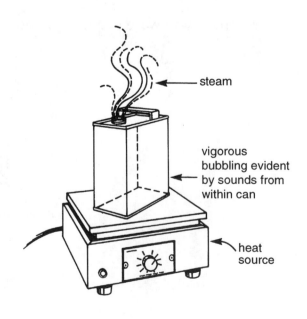

steam

vigorous bubbling evident by sounds from within can

heat source

3. Using a towel or oven mitts to prevent burns, steady the can with one hand and insert the rubber stopper into the opening of the can with the other hand.

4. IMMEDIATELY remove the can from the heat source and set it on a heat-resistant pad. If you leave the can on the heat source for any length of time after you seal it, you run the risk of exploding the can or, at the very least, shooting the stopper out with enough force to penetrate a tile ceiling.

5. If the water in the can was boiling vigorously, and if the can holds its airtight seal, it should slowly be crushed. Ask students to observe and describe what happens to the can as it slowly cools. Have them write their descriptions under 1 on the student page.

6. Now, have students brainstorm how they could get this result to happen faster. Have students write their predictions under 3 on the student page. Guide the discussion with specific questions so that students realize that cooling plays a role in crushing the can. ("Were we still heating the can at the time it started to crumple? Do you think this was a factor in what happened?") Therefore, if the rate of cooling is speeded up, the rate of crushing should also increase.

7. Repeat steps 1–4, but for step 5, hold the sealed can over a dishpan or sink and slowly pour cold water over it. The rate at which this can is crushed should be dramatically greater than the rate of crushing for the first can.

Safety Concerns

1. The cans will be boiling hot, so you will need something to protect your hands from burns as you steady the can to seal it or pour water over it.

2. After sealing, quickly remove the cans from the heat source. Be aware that the steam in the can will build up faster than expected. An exploding can could spray students with boiling water, and/or a forcefully expelled stopper could become a dangerous rubber "bullet." USE EXTREME CAUTION. Also, have a heat-resistant pad to set the first can on so as not to scorch or mar the desk or tabletop.

3. You should wear safety goggles at all times, and position students a safe distance from the can and heat source during the heating and sealing phases of this activity.

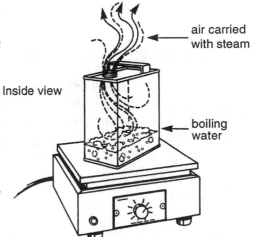

air carried
with steam

Inside view

boiling
water

Outcomes and Explanations

1. Before heating, the cans are filled with water and air. As you boil the water, some of it changes states from liquid to gaseous, becoming steam. As this steam flows out of the can, it carries some of the air in the can with it.

When the can is sealed, there is less air inside the can than outside the can, but the steam pressure built up by the boiling water keeps the can momentarily inflated. However, as the air in the can cools, it loses steam pressure, and the invisible "fingers" of the greater air pressure outside the can slowly crush the can.

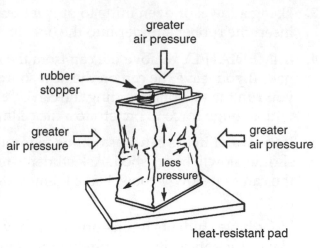

2. Discuss the effects of the changing air pressure, and have students write the explanation under 2 on the student page.

3. As previously discussed, pouring cold water over the second can will increase the cooling rate of the can. The cold water lowers the steam pressure in the second can faster than the air cooling did in the first can, allowing the outside air pressure to act more quickly on the second can. Discuss this difference with students, and have them write the explanation under 3 on the student page.

Application

Following the steps below, have students calculate how much air pressure (weight) is actually crushing the can.

1. Measure the dimensions of all six sides of the can. Determine the surface area of each side by multiplying the width times the length of that side. Add the surface area of all six sides. This should give you the total surface area of the can.

2. Multiply the total surface area of the can by approximately 15 pounds per square inch if you measured dimensions in inches. Multiply the total surface area of the can by 1 kilogram if you measured dimensions in centimeters. These numbers represent the approximate weight of air at sea level.

   A gallon can will have around 250 square inches of total surface area. That means you have an atmospheric pressure (weight) on the can of around 3,700 pounds. This is more than the weight of most automobiles. No wonder the can is crushed.

3. However, while the can is crushed, it isn't totally flattened. Shouldn't this much weight crush the entire can? Discuss with your students this apparent discrepancy in the effect of the air pressure on the can, and see what solutions they offer. (While the steam flowing out of the can removes some of the air in the can, all of the air inside the can is never removed. You do not have 3,700 pounds of pressure outside and none inside. The remaining air inside the can prevents the can from being totally flattened.)

Take Home

The threat of burns and the possibility of sealed cans exploding from overheating pose too great a risk for students to attempt this activity outside the classroom.

3 Uncle's Animal

Problem: *Can you save this animal?*

Your uncle brings you an unusual mammal about the size of a large house cat. He acquired this mammal during a recent trip around the world. The animal is so unusual that neither you nor your uncle knows exactly what mammal it is. Your uncle says he thinks the animal has become sick. It doesn't eat and no longer moves around very much. Your uncle is leaving for another trip, and you promise you will do your best to cure the sick animal. You have the following facts to work with:

1. You have a few days left before the animal becomes dangerously ill.

2. You can anesthetize the animal to examine it.

3. Expert help is not available.

4. The animal is common and well-known in some parts of the world.

Below is a list of actions you might take. Some actions are more important and should be done before others. Use the chart on the following page to rank from most important (1) to least important (14) the actions you would take. List your reasons for the rank you gave each action.

- Take the animal's temperature.
- See if the animal can swim.
- Read about the animal's natural habitat.
- Check the animal's mouth for infection.
- Ask your friends what to do.
- Make a list of the animal's physical characteristics.
- Read about the animal's food and water requirements.
- Examine the animal's waste droppings.
- Force-feed dog food to the animal.
- Adjust the temperature and light to match the animal's natural habitat.
- Use references to compare the animal's characteristics to those of other mammals.
- Observe the animal carefully.
- Collect a blood sample to be examined under the microscope.
- Offer the animal a drink of water.

Rank	Action Taken	Reason

3   For the Teacher

Objective

In this activity, students will use their powers of critical thinking and their own personal experiences to develop a plan of treatment for a sick, but unfamiliar, animal.

Materials Needed

- chart on student page showing the student's actions ranked in order of importance and the reasons for each ranking

Curiosity Hook

Display photos of some unusual animals.

Setup

1. Help students imagine themselves in the scenario given at the beginning of the student page.
2. Have students rank the actions they would take from 1 (most important) to 14 (least important) and give reasons for their rankings.
3. Students can work individually or in small groups.
4. Have students or groups share and defend their rankings. It may cause some consternation among your students, but there are no absolutely right or wrong answers. You are looking for soundness of reasoning, not correctness.

Safety Concerns

None

Outcomes and Explanations

Animal care experts rank the actions in the order shown on the next page.

Application

In this activity, students learn how to gather information to solve a problem and then how to evaluate and prioritize this information.

Take Home

Students might find it fun and interesting to share this activity with their parents, siblings, and friends to see how well they perform as animal care "experts."

Rank	Action Taken	Reason
1	Observe the animal.	Observe the animal's characteristics and behavior to help identify the animal and determinine if it shows any abnormalities.
2	Offer the animal water.	Water is basic to survival, can be easily offered, and is available while other actions are performed.
3	List its physical characteristics.	This list would provide necessary information for proper identification.
4	Compare it to other mammals.	This information would also help in proper identification.
5	Read about its natural habitat.	This information could be used to put the animal in a compatible environment.
6	Read about its food and water needs.	This information would help you to meet some of the animal's basic needs.
7	Adjust the temperature and light.	You would make the animal more comfortable and also meet some of its basic needs.
8	Take the animal's temperature.	You would gather information relating to the nature of its illness.
9	Check its mouth for infection.	You would gather information relating to the nature of its illness.
10	Examine the waste droppings.	You would gather information relating to the nature of its illness.
11	Collect a blood sample.	This is a more difficult task, but it might help to diagnose the illness.
12	Force feed the animal dog food.	This might satisfy the animal's food needs. However, it might also cause undo stress or even be harmful.
13	Ask your friends for advice.	Assuming your friends are not animal care experts, or are unfamiliar with this animal, this would be a waste of time.
14	See if it can swim.	This action would provide no useful information and could be quite stressful to the animal.

4   Apple I Liftoff

Problem: *Can you lift the apple off the table?*

Predict

1. Using the materials the teacher has laid out on the table, list as many ways as you can think of to lift the apple off the table. There are two rules:

   a. You have only the materials on the table to work with.

   b. You can handle, arrange, and assemble any or all of the materials in any way you wish, but when it comes time to move the apple off the table, you may only touch the spool.

Conclude

2. What did the teacher do to solve this problem, and how does the teacher's method work?

4   For the Teacher

Objective

In this activity, students will use their powers of observation, critical thinking, and problem solving to meet the challenge of lifting the apple off the table.

Materials Needed

- 3-foot piece of heavy string
- 1 large spool
- 1 apple
- 1 one-hole rubber stopper
  (You might borrow this from your chemistry department.)
- a large paper clip
- a pencil or wooden dowel to punch a hole through the apple
- safety goggles

(To save time, you may want to have a completed assembly ready and kept out of sight until you need it.)

Curiosity Hook

As students enter the classroom, display the unassembled materials on a table and toss the apple in the air.

Setup

1. Challenge students to think of as many ways as possible to lift the apple off the table. Solutions to the problem must follow the rules given on the student page. Have students write and/or diagram their responses to this challenge under 1 on the student page.

2. One fairly complex method is to assemble the materials as follows:

   a. Tie one end of the string to the one-hole rubber stopper.

   b. Pass the other end of the string through the large spool.

   c. Use a pencil or wooden dowel to punch a hole through the apple.

   d. Push the end of the string through the apple and secure the end of the string with the paper clip. The assembled apparatus should look like the illustration on the next page.

3. Grab the spool and slide it up the string until the rubber stopper rests on top of the spool.

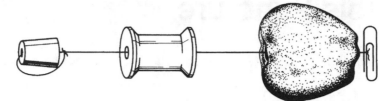

4. Now, start to swing the rubber stopper around by moving the spool in a circle. The rubber stopper will begin to move outward, pulling the string up through the spool and lifting the apple off the table.

Safety Concerns

Wear safety goggles as you twirl this apparatus. Keep students back a safe distance. A spinning rubber stopper could give a painful whack in the eye.

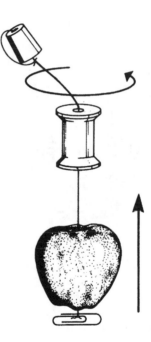

Outcomes and Explanations

1. Discuss this challenge with your class. Accept and list any and all reasonable student solutions. You might actually test some of the student solutions to this problem.

2. The spinning rubber stopper produces a centrifugal force outward. As the stopper swings in an ever-widening circle, it moves even faster, producing enough force to pull on the string and lift the apple off the table.

3. Discuss this result with students and have them write the explanation under 2 on the student page.

Application

Ask students where they may have seen or experienced such movements and forces. Examples of such movements and forces can be found in certain carnival rides when they may have felt themselves moved to the outside of their seats as a car went fast around a curve, and in a clothes washer when clothes flatten out as they spin in the drum.

Take Home

If students have the materials to construct a spinning-apple system at home, encourage them to do so. Such a system would be an excellent way for them to amaze and teach their parents, siblings, and friends. Caution them to keep spectators back a safe distance and urge them to wear eye protection when attempting this activity.

5 | Double Puncture

Problem: *If a container has two holes in it, will water leak out both holes?*

Observe

1. Observe the system you see before you and, in the space below, describe this system.

Predict

2. Predict what will happen when the nails are removed.

    A. top nail only

    B. bottom nail only

    C. both nails

Conclude

3. What did happen? Did the position of the nail make a difference? Did the number of nails removed make a difference?

A. top nail only

B. bottom nail only

C. both nails

5   For the Teacher

Objective

In this activity, students will use their powers of observation and problem solving to learn that all holes do not necessarily leak.

Materials Needed

- 1 plastic soda container
  (Size doesn't matter, but the larger the container, the easier it is for students to see the results.)

- 1 screw-on lid or rubber stopper to seal the container

- 2 small push pins or nails
  (Make sure the nails are smooth, with no burrs. The smoother the nails, the better they seal the holes they are in.)

- enough water to fill the container

- a sink or dishpan for the container to drain into

Curiosity Hook

Have the container full of water with the nails in it sitting where students can see it when they come into the classroom

Setup

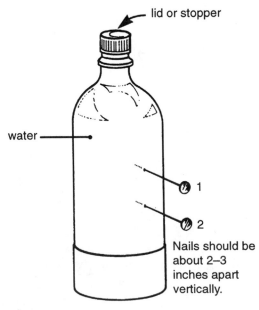

lid or stopper

water

1

2

Nails should be about 2–3 inches apart vertically.

1. The double-punctured container should be assembled as shown.

2. Ask students to hypothesize what will happen when only one nail is removed at a time and then when both nails are removed. Have students write their predictions under 2 on the student pages.

3. Remove the top nail only. Water will not leak out. (A small amount may initially spurt out and then stop.)

4. Remove the bottom nail only. Water will not leak out. (A small amount may spurt out and then stop.)

5. Now, remove both nails. Air will bubble into the container through the top hole and water will leak out of the container through the bottom hole.

6. Have students write these results under 3 on the student pages.

Safety Concerns

Use common sense with pins and nails.

Outcomes and Explanations

1. As long as the container is sealed, water will not leak out while there is only one hole. Why? When you sealed the container you eliminated air pressure on top of the water. Now you have only the weight (pressure) of the water pushing inside the hole and air pressure pushing outside the hole.

   The forces on either side of the hole are balanced, so nothing moves in or out.

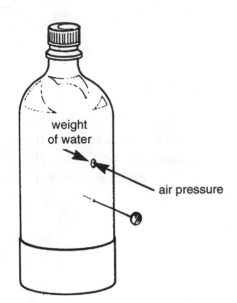

weight
of water

air pressure

2. When both nails are removed, the forces become unbalanced, and air pushes into the top hole while water pushes out the bottom hole.

3. Discuss the changes in pressure with students and have them write the explanation under 3 on the student pages.

Application

Challenge students to investigate the following problems:

1. Does the orientation of the nail holes make a difference? Will it make a difference if you put the nail holes side by side (horizontal orientation) rather than one above the other (vertical orientation)?

2. Does the size of the container make a difference?

3. Does the orientation of the container make a difference? When the container is sitting upright (vertical) and both nails are removed, air comes in the top hole, and water leaks out the bottom hole. What will happen if you remove both nails, pick up the container, and hold the container horizontally (as shown on the next page)?

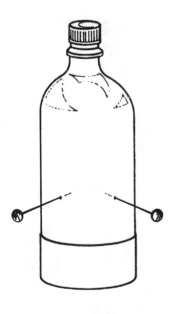

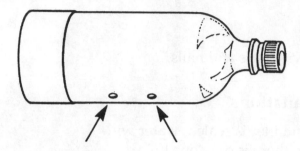

Take Home

This activity can be easily and safely done by students outside the classroom. Encourage students to investigate on their own the challenge problems listed under the Application section. This activity can get messy, so caution them to conduct their experiments in a location where spilling water will not be a problem.

6 Super Boiling

Problem: *Do you really understand all you know about boiling?*

Observe

1. The teacher is going to boil water in an open flask. How will you know when the water is boiling? Give two indicators of boiling.

    A.

    B.

Predict

2. If the teacher removes the flask from the heat or the heat from the flask, will the water stop boiling? Why?

3. Will adding anything to the water make a difference? The teacher will bring the water to a boil again, but this time he/she will add some salt. Predict what will happen.

Conclude

4. What did happen and why?

6   For the Teacher

Objective

In this activity, students will use their powers of observation, prediction, and critical thinking to investigate the less-than-obvious facets of the process of boiling.

Materials Needed

- 1 flask
  (Size doesn't matter, but the larger the flask, the easier it will be for students to see the results.)

- a source of heat
  (If you have access to natural gas outlets, use a Bunsen burner. Otherwise, a butane burner or a hot plate will suffice. In a pinch, an alcohol burner will work. However, it will take a long time to boil water using only an alcohol burner.)

- 1 metal ring stand and wire screen to set the flask on while boiling
  (You may be able to borrow a ring stand and wire screen from your chemistry department.)

- 1 chemical thermometer (glass only, not aluminum-encased)
  (This may be borrowed from your chemistry department.)

- enough water to fill the flask about half full

- a container of ice cubes or crushed ice

- a shaker or container of table salt

- safety goggles

- kitchen hot pads, heat-resistant gloves, or some other device for handling the flask of boiling water

- food coloring
  (This is optional and may be added to the water in the flask for visual effect.)

Curiosity Hook

Boil a flask of colored water where students can see it as they enter the classroom.

Setup

1. Fill the flask about half full of water and set it on the stand.

2. Begin heating the water. While waiting for the water to boil, have students discuss and answer question 1 on the student page.

3. Once the water begins to boil, have the students note the bubbling action of the water. Then put the thermometer in the boiling water and show that the temperature is indeed 100°C or 212°F.

4. Ask students to predict what will happen if you remove the heat source or remove the flask from the heat source. Have students write their predictions under 2 on the student page. Remove the heat from the water.

5. Bring water to a boil again. Now have students predict what will happen when you add a small amount of table salt to the boiling water. Have students write their predictions under 3 on the student page.

Safety Concerns

1. You should wear safety goggles while conducting this demonstration.

2. Position students a safe distance from the heat source and boiling water.

3. Have some types of heat-resistant materials available—kitchen hot pads, special gloves, etc.—for handling a boiling hot flask.

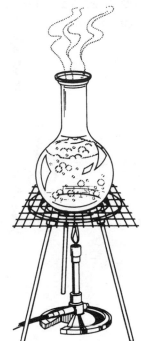

Outcomes and Explanations

1. When answering question 1 on the student page, students usually come up with two main signs of boiling:

    a. bubbling

    b. temperature
        (For your reference, the boiling point of water at sea level is 212° F or 100° C.)

2. When the heat is removed from the boiling beaker, the temperature drops and boiling stops. Obviously, heat is a major factor in boiling.

3. As soon as salt is added to the boiling water, the temperature drops and the water stops boiling. This is because the salt has a lower temperature and dissolving it in the water absorbed some of the heat.

    With the salt in the water, the water molecules adhere not only to each other but also to the salt ions, making it harder to transfer the water into a vapor (steam) state. Therefore, what we think of as the boiling temperature of water must apply only to pure water.

4. Will adding even more salt elevate the boiling temperature of the water even further? Extend this investigation by having students consider this question (and yes, the more salt, the higher the boiling temperature of the water).

Application

1. Challenge students to think of ways to make practical use of this information. One practical application is adding alcohol antifreeze (salt would be too corrosive) to the

radiators of our automobiles to both raise the boiling temperature and lower the freezing temperature of the water in the radiator.

Take Home

The safety hazard presented by hot temperatures and boiling water, and the necessity of having specialized equipment, make it impractical for students to duplicate this demonstration outside the classroom.

7 Bouncy, Bouncy

Problem: *How high will the smaller ball bounce?*

Predict

1. In the space below, predict what will happen when the two balls are dropped together with the smaller ball on top.

Conclude

2. What did happen, and why did it happen that way?

7 For the Teacher

Objective

In this activity, students will use their powers of observation, critical thinking, and problem solving in an attempt to understand why momentum causes the small ball to fly up much faster than it falls down.

Materials Needed

- 2 balls of unequal size
(A tennis ball and a basketball will work nicely.)

Curiosity Hook

Bounce the large ball as students enter the classroom.

Setup

1. Start by having students observe you holding the two balls, with the small ball on top of the larger ball.

2. Ask students to predict what will happen to this system when you drop both balls at the same time with the smaller ball on top. Have students write their predictions under 1 on the student page.

3. When dropped in this configuration, the smaller ball flies upward with a much greater velocity than it had in falling down. It also flies much higher than the original height from which it was dropped. Have students record this result under 2 on the student page.

Safety Concerns

1. The smaller ball will fly off the larger ball with surprising speed and force and often at unexpected angles, so position students well back from the drop zone.

2. Consider doing this activity outside or in a large open area like a gymnasium to minimize the possibility of something in your classroom getting broken by flying balls.

Outcomes and Explanations

1. Why does the small ball fly upward faster than it came down and higher than the height it was dropped from? In one word—momentum.

2. As the small ball falls and nears the floor, it collides with the upward-rising large ball. This collision results in the momentum of the larger ball being transferred to the smaller ball.

3. Momentum is a function of mass or size. The larger ball has much more momentum than the small ball, and when this large amount of momentum is transferred to a smaller object (the small ball), the smaller object literally rockets upward.

4. Discuss momentum with students, and have students write the explanation under 2 on the student page.

Application

Challenge students to consider the following problems:

1. What if we drop two balls of the same size? (There is little or no added velocity or height to the ball on top.)

2. What if we drop more than two balls? (Say you drop four balls of different sizes in a configuration like that illustrated. Ball 1 bounces up with a velocity close to what it had in falling, but ball 2 bounces up much faster, ball 3 even faster, and ball 4 will shoot upward at a tremendous comparative speed to over five times the height from which it was dropped.)

Take Home

This is an easy activity for students to try at home. Encourage them to try the challenge problems given under the Application section.

8 Launch Those Pennies!

Problem: *Do all objects fall at the same rate, or does gravity play favorites?*

Predict

1. Carefully observe the penny-projectile launcher that your teacher has prepared. The teacher is going to flick the card. This will cause one penny to be thrown away from the launcher, while the other penny will fall straight down. Predict which penny will hit the floor first. Use sound rather than sight, and listen for the click of the pennies hitting the floor.

Conclude

2. Which penny did hit the floor first, and why did it happen that way?

Predict

3. Let's try something different. This time the teacher will replace one of the pennies with a larger and heavier coin. Predict which object will hit the floor first when the card is flicked. Again, listen for the clicks.

Conclude

4. Which object did hit the floor first, and why did it happen that way? What conclusions can you draw from all of this?

8 For the Teacher

Objective

In this activity, students will learn that gravity does not play favorites.

Materials Needed

- 1 note card, 3" × 5"
- 3 coins
  (2 pennies and 1 quarter)

Curiosity Hook

Toss and catch a coin in the air as students come into the classroom.

Setup

1. To make a penny-projectile launcher, fold the note card in half then fold each side outward one third from the edge.

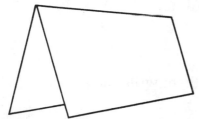 

2. Grasp one corner of the card with your thumb and forefinger. (You may prefer to grasp a corner of the center ridge.) Place the pennies on opposite sides of the center ridge.

3. Hold the loaded penny-projectile launcher level and at about chest height. Ask students to predict which penny will hit the floor first when you flick the card. Have students write their predictions under 1 on the student page.

4. Flick the center ridge of the card with the middle finger of your free hand.

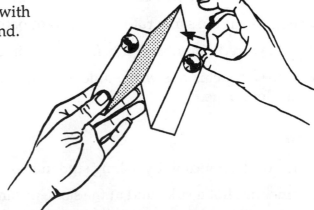

5. Have students listen rather than look for results. They should hear one click, as both pennies will hit the floor at the same time.

6. Now, repeat this experiment, but use a penny and something larger and heavier, like a quarter. Have students predict which coin will hit the floor first. Have students write their predictions under 3 on the student page.

7. Again, they should hear only one click, as both objects hit the floor at the same time.

Safety Concerns

Use common sense.

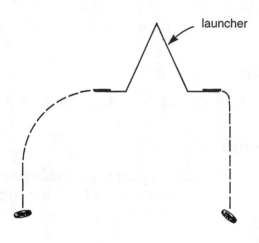

launcher

Outcomes and Explanations

1. Most students will predict that the penny falling straight down will hit the floor first. However, both pennies hit the floor at the same time (one click). Why? The moment both pennies are launched, gravity begins to pull them down with the same speed. Thus, they both accelerate downward at the same rate, one falling in a curved path and one falling straight, but both arriving at the floor at the same instant.

    Discuss with students this effect of gravity, and have them write the explanation under 2 on the student page.

2. When a penny and a larger and/or heavier object are launched, students will usually predict that the larger and/or heavier object will hit the floor first. Again, they both hit the floor at the same time. Actually, objects of different sizes and weights will only fall at exactly the same rate in a vacuum. In our case, we disregarded air resistance because of the relatively short distance of the fall. Discuss the effect of weight on the falling rate of objects, and have students write the explanation under 4 on the student page.

3. Students should draw two conclusions from this activity:

    a. Objects fall at the same rate regardless of path.

    b. Objects with different weights fall at the same rate.

    Discuss these conclusions with students, and have them write these conclusions under 4 on the student page.

Application

1. Check for understanding by asking students to contemplate these questions:

    a. If someone shot a rifle, and at the same instant the bullet left the barrel, a bullet was dropped straight down from the same height as the barrel of the rifle, which bullet would hit the ground first? (As unbelievable as it sounds to students, they would both strike the ground at the same time. Remember our first conclusion: objects fall at the same rate regardless of path.)

b. A glass tube is held horizontally, and a penny and a feather are put into one end of the tube. The tube is sealed, and all the air inside is pumped out (in other words, there is a vacuum inside the tube). The tube is then turned vertically, and both objects begin to fall. Which will hit the bottom of the tube first? (Again, both will hit at the same time.)

2. Have students construct their own projectile launchers, and challenge them to investigate the following problem: Does the shape of the object determine which object falls faster? Have students launch objects of different shapes and see if shape makes a difference in the rate of fall. Encourage good scientific method here. Both objects should be different shapes but the same approximate weight. Only test one variable at a time.

Take Home

Encourage students to conduct this activity outside the classroom and use it amaze and teach their parents, siblings, and friends.

9 The Mysteries of Möbius Strips

Problem: *What mysteries are revealed when you turn ordinary loops of paper into variations of the Möbius strip?*

Predict

1. In the space below, predict what will happen when a loop of paper is cut down the middle.

Conclude

2. What did happen when a loop of paper was cut down the middle, and why did it happen that way?

Predict

3. In the space below, predict what will happen when we do the same thing but give the paper loop a half twist first.

Conclude

4. What did happen when the paper loop with a half twist was cut down the middle?

Predict

5. Baffled yet? In the space below, predict what will happen when we do the same thing but we give the paper loop a full twist.

Conclude

6. What did happen when a loop of paper with a full twist was cut down the middle?

9 For the Teacher

Objective

In this activity, students will learn that things are not always as they appear, at least not with the strange paper loop called a Möbius strip.

Materials Needed

- 1 roll of adding machine paper
- scissors
- 1 roll of cellophane tape

Curiosity Hook

Twirl a small loop of paper on your finger as students come into the classroom.

Setup

1. Pull about 4 feet (48 inches) of paper off the roll and cut it. This gives a loop about 2 feet (23–24 inches) long. The exact size of the loops is not critical, but make them large enough so that students can easily see what is happening as you cut them. For comparison purposes, regulate variables and make each subsequent loop approximately the same size as the first one.

2. Now, form the paper into an untwisted loop, slightly overlap the ends, and tape. On all loops, tape the entire edge down.

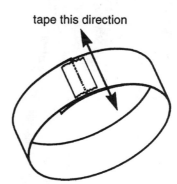

tape this direction

3. Have students predict what will happen when you cut the loop down the middle. Have students write their predictions under 1 on the student pages. Cut the loop down the middle. As most students will probably predict, you end up with two loops the same size as the original. Have students write the results under 2 on the student pages.

4. What will happen if you start twisting the paper loops? Form a loop with a half twist. Have students predict what will happen if you cut down the middle of a loop with a half twist in it, and have them write their predictions under 3 on the student pages. Cut down the middle of the loop with a half twist. When done, you should

have a single loop, but it will be twice as long as the original. This strip is called a Möbius strip. Have students write the results under 4 on the student pages.

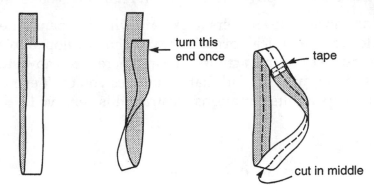

5. Now, form a loop with a full twist. Have students predict what will happen if you cut down the middle of a loop with a full twist in it. Have students write their predictions under 5 on the student pages. When you cut down the middle of a loop with a full twist in it, you get two loops with a full twist the same length as the original but linked together. Again, this variation of the Möbius strip baffles and amazes. Have students write the results under 6 on the student pages.

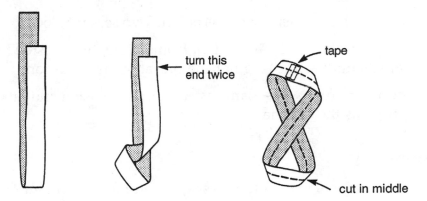

Safety Concerns

Use common sense with scissors.

Outcomes and Explanations

1. When cut down the middle, the loop with no twists becomes two separate loops the same size as the original. You obviously cut the original loop in two pieces.

2. However, when you start twisting the loops and then cutting, the results are anything but obvious. The twisted loops of paper, such as those you have formed, are variations of the Möbius strip. The Möbius strip takes its name from August Ferdinand Möbius (1790–1868), a German astronomer and mathematician. Möbius helped establish a study in geometry called topology. Topology deals with geometrical figures that are pulled or twisted out of shape in various ways.

3. The amazing thing about a Möbius strip is that it only has one side; the inside is the

outside! You can demonstrate and confirm this for your students (and yourself) by making a loop with a half twist and coloring one side of the loop. When you are done, the entire loop will be colored. It really does have only one side.

4. There are no simple answers to explain what happens when you cut Möbius strips. Complex mathematical formulas have been developed to explain them, but that is not the purpose of this activity. The aim here is to show students that things are not always as they appear, and that sometimes you can't even trust your own eyes. The Möbius strip and its variations illustrate this very well.

Application

Have students investigate the following:

1. What happens when you cut a loop with one and a half twists (3 turns) down the middle? (You will get one large ring with a knot in it!)

2. Does it make a difference where you cut the loop? Form a new set of loops—no twist, half twist, and full twist. Have students predict what will happen if you cut each of these loops about 1/3 from one edge rather than down the middle. The results should be as follows:

   a. loop with no twist—same as before, two separate loops

   b. loop with half twist—different from before, two twisted but linked loops, one the same length as the original and one about twice as long

   c. loop with full twist—same as before, two twisted and linked loops both the same length as the original

Take Home

Encourage students to cut old newspapers and form them into variations of the Möbius strip to amaze and teach their parents, siblings, and friends.

10 | Defying Gravity

Problem: *Is it possible to invent a device that defies gravity and rolls uphill?*

Observe

1. In the space below, accurately describe the "gravity-defying" device and the ramp system the teacher has prepared.

Predict

2. Predict what will happen when the teacher places the "gravity-defying" device on the ramp.

Conclude

3. Did the device defy gravity and roll uphill? Explain why or why not.

10 For the Teacher

Objective

In this activity, students will use their powers of observation, prediction, critical thinking, and problem solving to see the truth behind the illusion of a device that appears to roll uphill.

Materials Needed

- 2 meter sticks or yardsticks
- 2 glass funnels
  (Exact size is not critical. The measurements given in the setup explanation are based on a funnel with a top opening of 4 in. [10 cm]. Plastic funnels will work, but not as well. They are so lightweight that you may have to nudge them to get them to roll.)
- assorted books as supports for the meter sticks
- masking tape

Curiosity Hook

Roll the "gravity-defying" device (two taped funnels) along the top of a desk as the students enter the classroom.

Setup

1. Assemble a "gravity-defying" device by taping two glass funnels together.

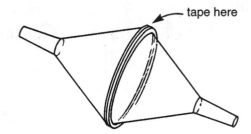
tape here

2. Assemble a ramp system similar to that shown below. Measurements are approximations.

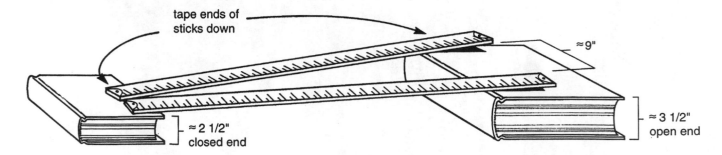
tape ends of sticks down
≈9"
≈2 1/2" closed end
≈3 1/2" open end

3. Make it a point to show students that the open end of the ramp system is higher than the closed end.

4. Ask students to predict what will happen when the "gravity-defying" device is placed on the closed end of the ramp system. Have students write their predictions under 2 on the student page.

5. Now, place the taped funnels onto the closed end of the ramp system. The funnels will <u>appear</u> to roll uphill. Have students write this result under 3 on the student page.

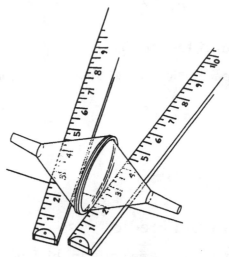

Safety Concerns

None

Outcomes and Explanations

Actually, the funnel system doesn't roll upward at all. If you observe the system from the side as the funnels roll along the ramp, you'll see that the taped edge of the funnels actually drops down into the open space between the two rulers. The illusion works because the sides of the funnels are sloped, allowing the taped edge of the funnels to drop down while making it appear as if the sides are rolling upward. It is the ramp that goes up, not the funnels. Have students write the explanation under 3 on the student page.

Application

Challenge students to investigate the following problems:

1. What will happen if we change the shape of the ramp? (If the meter sticks are placed parallel to each other, the funnels will not move.)

2. What will happen if we change the height of the ramp supports? (If both ends of the ramp system are made the same height, the funnels will not move.)

Take Home

If students have access to funnels and sticks, encourage them to use this activity to amaze and teach their parents, siblings, and friends.

11 Leap Ball

Challenge: *Can you move a Ping Pong ball from one cup to another without directly touching it?*

Observe and Describe

1. In the space below, observe and describe/diagram the system the teacher has prepared:

Brainstorm and Predict

2. List as many ways as you can think of to move the Ping Pong ball from one cup to the other. The only rule is that you cannot touch the Ping Pong ball directly.

Conclude

3. Did the teacher solve the challenge? If so, explain how the teacher's method worked. If not, explain why not.

11 For the Teacher

Objective

In this activity, students will be challenged to use their skills of observing/describing, predicting, analyzing, and concluding to solve the problem of moving a Ping Pong ball from one cup to another.

Materials Needed

- 2 8-oz (250 ml) Styrofoam cups
- 1 Ping Pong ball

Curiosity Hook

Toss the Ping Pong ball in the air and catch it as students enter the room.

Setup

1. Place the cups side by side in a location where they can be easily observed by all students. Show them the Ping Pong ball, and then drop the ball into one of the cups.

2. Students should now describe this setup under 1 on the student page. Remind them that the easiest and most accurate way to describe something is to make a labeled diagram of it.

3. Now issue the challenge: Can you move the Ping Pong ball from one cup to the other? Inform students that there is only one rule—they cannot directly touch the ball. How many ways can students come up with to solve this problem? Give them a set amount of time to brainstorm and have them write their solutions under 2 on the student page.

4. If time permits, let students try some of their solutions to see if they work. The simplest solution within the rules would be to just dump the ball from one cup into the other. Demonstrate this dramatic solution to the problem: while holding both cups firmly, blow with short, hard puffs into the far side of the cup containing the ball. The ball will literally leap out and land in the other cup. As usual, you will want to practice ahead of time so you get some feel for how hard and at what angle to blow in order to get the ball to fly out and into the other cup.

Safety Concerns

None

Outcomes and Explanations

1. The ball leaps out of the cup because an updraft of air pressure is created by blowing

into the cup at an angle. The air flowing over the top of the cup creates a low-pressure area above the ball. The air in the lower part of the cup is now higher in pressure than the air above the cup and pushes the ball out of the cup quite forcefully.

2. Help students figure out how and why your solution worked and have them write the explanation under 3 on the student page.

Application

The ball leaps out (in reality is pushed out) of the cup because of Bernoulli's principle of pressure and velocity. Applications of this principle are found in spray-painting guns, spray bottles, and pressurized spray cans.

Take Home

If students have the necessary materials at home, encourage them to amaze and educate their family and friends by demonstrating this principle to them.

12 Stop That Leak!

Problem: *Can you stop the leak?*

Observe

1. In the space below, accurately describe the situation before you.

Predict

2. List as many ways as you can think of to stop the leak. There is only one rule: You cannot put anything into or over the hole where water is leaking out.

Conclude

3. What did the teacher do to solve the problem, and how does the teacher's method work?

12 For the Teacher

Objective

In this activity, students will use their powers of observation and problem solving in an attempt to stop the leak.

Materials Needed

- 1 plastic soda container
  (The exact size is not critical but the larger the better.)

- 1 one-hole rubber stopper that fits snugly into the opening of the soda container
  (If a stopper is not available, use the cap from the container)

- an awl or small nail to make a hole in the container

- enough water to fill the container several times

- food coloring
  (This is optional and could be used to color the water in the container for visual effect.)

Curiosity Hook

Have water leaking from a small hole in the container when students come into the classroom.

Setup

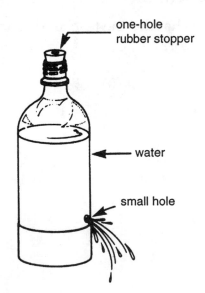

one-hole
rubber stopper

water

small hole

1. Prepare the container as illustrated. (If using the cap instead of the stopper, have the top of the container open.)

2. Have students carefully observe the leaking container and then write their descriptions under 1 on the student page.

3. Challenge students to stop the leak according to the rule stated under 2 on the student page. Have students write their solutions to the problem. Urge them to make their solutions as simple and practical as possible.

Safety Concerns

Use common sense when poking holes in the container.

Outcomes and Explanations

air pressure

1. Ask students to share their solutions to the problem with the rest of the class. Accept all reasonable solutions. Try as many of the student solutions as practical to verify that they will actually work.

2. My solution to the problem is to put my finger tightly over the hole in the rubber stopper. (An alternative solution is to put the cap back on the container.) Be aware that the water flow out the hole will not stop immediately but will gradually cease. Have students write this solution under 3 on the student page.

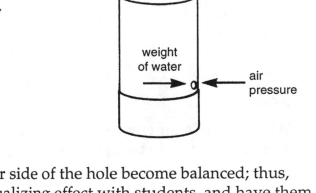

+
weight
of water

air
pressure

3. Why does putting your finger over the hole in the stopper stop the leak? The bottle leaks because there are forces working against each other. Air pressure on top of the water and the weight of the water (gravity) are pushing on the inside of the hole against air pressure pushing outside of the hole.

4. When you seal the bottle by putting your finger over the hole in the stopper, you eliminate the air pressure pushing down on the top of the water. Now, you have only the weight of the water pushing on the inside of the hole and air pressure pushing on the outside of the hole.

weight
of water

air
pressure

By closing the hole in the stopper or capping the container, the forces on either side of the hole become balanced; thus, nothing moves in or out. Discuss this equalizing effect with students, and have them write the explanation under 3 on the student page.

Application

This activity explains why we punch two holes in any can of liquid we are trying to pour. One hole allows air to enter while the liquid pours smoothly out the other hole.

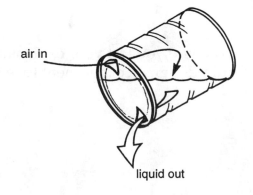

air in

liquid out

Take Home

Encourage students to challenge their parents and friends to solve this leaky problem.

13 | Bubble, Bubble

Problem: *What is causing the difference in "fizz rate" between the two containers?*

Observe

1. Observe and describe what happens when the teacher drops a seltzer tablet into each container.

   Container A

   Container B

2. The seltzer tablets have different "fizz rates." That is, they do not react at the same speed in both containers. What difference between the two containers could account for this difference in "fizz rate"? In the space below, list all the differences you can observe between the two containers.

Conclude

3. Which of these differences do you think accounts for the difference in "fizz rate"? Write your explanation in the space below.

4. How could you test your explanation (hypothesis)? In the space below, explain how you would design an experiment to test your hypothesis.

13 For the Teacher

Objective

In this activity, students will use their powers of observation, critical thinking, and experimental design to analyze variables and determine why the rate of bubbling (reaction rate) is different between two containers.

Materials Needed

- 2 clear glass or plastic containers of different shapes and volumes (Different size beakers and flasks work well.)

- hot and cold water

- 2 different colors of food coloring

- 2 different brands of seltzer tablets (Use two brands that have the same or similar reaction rates.)

Curiosity Hook

Have a seltzer tablet fizzing in a container of colored water where students can see it as they come into the classroom.

Setup

1. Place a container of water in a refrigerator the day before you plan on conducting this activity.

2. On the day scheduled for this activity, prepare as follows:

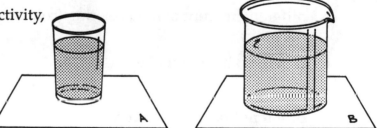

   a. Set one of the clear glass or plastic containers on a piece of paper marked A and the other container on a paper marked B. These are the demonstration containers.

   b. Just before students come into the classroom, prepare two stock solutions in containers other than the demonstration containers. Designate one container as Stock Solution A. Make this solution by mixing hot tap water with a little food coloring (say red). The other container will be designated as Stock Solution B. Make this solution by mixing the cold water from the refrigerator with a little food coloring (say blue). Use different amounts of food coloring so one stock solution is obviously darker than the other. The color difference adds another variable that students must contend with. Students should not see the stock solutions until they are needed.

3. Once students are seated and ready to proceed, bring both stock solutions out of hiding. Pour Stock Solution A into Demonstration Container A and Stock Solution B into Demonstration Container B. Put different amounts of liquid in each demonstration container. This adds yet another variable.

4. Drop a different brand of seltzer tablet into each demonstration container. The seltzer tablet in container A should react faster than the one in container B. Have students describe what is happening in each demonstration container under 1 on the student page.

5. Now, have students list all the differences (variables) they can observe between the two demonstration containers. Have students list the differences under 2 on the student page.

6. Ask students to pick the difference (variable) they feel is responsible for the difference in "fizz rate" (reaction rate) between the two demonstration containers. Have students write their predictions under 3 on the student page.

7. Challenge students to design an experiment to test their prediction. Have students describe their experimental design under 4 on the student page.

Safety Concerns

Use common sense.

Outcomes and Explanations

1. Students should determine that the following differences exist between the two demonstration containers:

    a. different color liquid in each container

    b. different amount of food coloring (color in one container is visibly darker than in the other)

    c. different shape containers

    d. different size (volume) containers

    e. different brand of seltzer tablets

    f. different temperature of liquid in each container

    g. different amount of liquid in each container

    Discuss these differences with students, and have them list these variables under 2 on the student page.

2. How could you test to see which of the variables is causing the difference in reaction rate? Simple. Keep all other variables the same except the one you want to test. Suppose you believe that the difference in color is causing the difference in reaction rate. Set up two containers that are identical in all variables except color of liquid. If, when you do this, the seltzer tablets do not react at the same speed, you have proven that the color difference is causing the reaction rate difference. However, if the tablets

do react at the same speed, color difference has nothing to do with reaction rate difference.

Discuss experimental design with students. Make sure they are clear on this procedure. Have students share their experimental designs with the rest of the class and then have the class evaluate each design.

3. Share the answer with students. The correct variable is temperature. The reaction ("fizz rate") proceeds faster in Demonstration Container A because the liquid in that container is warmer.

Application

Have students list as many examples as they can think of in everyday life where we try to speed up reactions by heating something. For example, to make bread, yeast is usually added to warm water to activate it and then added to the dough mixture to cause the dough to rise.

Take Home

If students have access to the necessary materials, encourage them to try this activity at home.

14 Stop Rolling

Problem: *Will a jar of sand roll as far as a jar of water?*

Predict

1. Predict what will happen when a jar half full of water is released at the top of a ramp. Write your prediction in the space below.

Conclude

2. What did happen when the jar was released, and why did it happen that way?

Predict

3. Now, predict what will happen when a jar half full of sand is released at the top of a ramp. Write your prediction in the space below.

Conclude

4. What did happen when the jar was released, and why did it happen that way?

14 For the Teacher

Objective

In this activity, students will use their powers of observation and critical thinking to determine why a jar half full of water rolls but a jar half full of sand does not.

Materials Needed

- 2 straight-sided clear plastic or glass jars with screw caps
  (The size of the jar is not critical, but both jars should be the same size.)

- enough water to fill one jar about half full

- enough sand to fill one jar about half full

- 1 board about 3 to 4 feet long to serve as a ramp

- books or bricks to support one end of the ramp

Curiosity Hook

Have the ramp set up on a table, and roll a small ball or some other object down the inclined plane as students come into the room.

Setup

1. Fill one jar about half full of water and the other jar about half full of sand. Screw the lids on both jars.

2. Use the board and supports (books, bricks, etc.) to make a ramp. Make the upper end of the ramp only about 2–3 inches high. You don't want jars skidding down a ramp that is too steep instead of rolling.

3. Now, hold the jar with water in it on the upper end of the ramp.

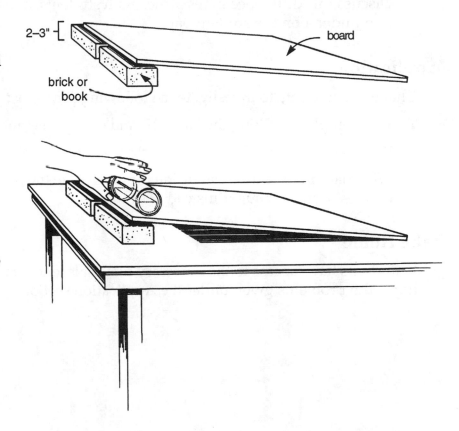

Ask students to predict what will happen when you release the jar. Have students write their predictions under 1 on the student page.

4. Release the jar half full of water. This jar should roll freely down the ramp. Have students write this result under 2 on the student page.

5. Now, hold the jar half full of sand on the upper end of the ramp, and ask students to predict what will happen when you release it. Have students write their predictions under 3 on the student page.

6. Release the jar half full of sand. This jar should not move or should move very little. Have students write this result under 4 on the student page.

Safety Concerns

Use caution when rolling the glass jars; they are heavy and could break.

Outcomes and Explanations

1. Friction is the key to why one jar rolls and the other does not.

2. Water offers very little resistance (friction) to the sides of the jar as the jar turns (rolls), so the jar rolls more or less freely. Discuss this effect with students, and have them write the explanation under 2 on the student page.

3. Sand, however, offers a great deal of resistance to the sides of the jar. So much resistance (friction), in fact, that the jar cannot turn and so cannot roll down the ramp. Discuss this difference in resistance with students, and have them write the explanation under 4 on the student page.

Application

Challenge students to investigate the following problems:

1. Will completely filling the jar with sand change the outcome?

2. Will changing the shape of the jar change the outcome?

3. Will changing the material inside the jar change the outcome? Try using flour, molasses, and/or sugar instead of sand.

Take Home

If students have access to the necessary materials, encourage them to investigate at home the problems given under the Application section.

15   The Mysterious Rising Water

Problem: *What causes the water to rise in the small beaker?*

Observe

1. In the space below, accurately describe the device you see before you.

Predict

2. The teacher will now begin to boil the water. What happens to the small beaker as the water boils? What do the bubbles in boiling water consist of?

Conclude

3. The teacher will now stop heating the large beaker and allow the device to cool. Describe what happens in both beakers as the water cools. Why does it happen that way?

15 For the Teacher

Objective

In this activity, students will use their powers of observation, critical thinking, and problem solving to understand why the water moves mysteriously into the small beaker.

Materials Needed

- a 600-ml beaker
- a 100-ml beaker
- a hot plate or a burner and stand
- about 120 ml of water
- food coloring
- safety goggles
- hot pads and heat-resistant pad

Curiosity Hook

Have the device set up where students can see it.

Setup

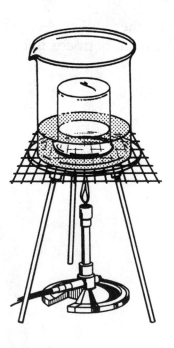

1. Put some food coloring into approximately 120 ml of water, and pour the colored water into the large beaker.

2. Invert the small beaker and put it down into the water in the large beaker.

3. Set the beakers on a hot plate or stand.

4. Have students carefully observe the device and then accurately describe it under 1 on the student page.

5. Heat the water until you have a vigorous boiling action. Let the water boil vigorously for several minutes. As the water in the large beaker boils, the small beaker will rock back and forth or bob around. Do not let the small beaker tip over during boiling.

6. Now, stop boiling. If using a burner and stand,

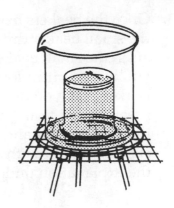

simply turn off the burner. If using a hot plate, re-move the beakers from the hot plate and set them on a heat-resistant surface.

7. As the water cools, it will leave the large beaker and move up into the small beaker. Have students de-scribe this under 3 on the student page.

Safety Concerns

1. You should wear safety goggles at all times, and position students a safe distance from the beakers and the heat source.

2. You will need something to steady the small beaker during boiling so it doesn't tip over. Do not use your fingers as the steam from the boiling water will be very hot. The eraser end of an ordinary pencil works well for this.

3. If using a hot plate, you will need something to protect your hands when removing the hot beakers from the burner and a heat-resistant pad to set the hot beakers on.

Outcomes and Explanations

1. Before you begin to boil the water, there is equal air pressure inside and outside the small beaker.

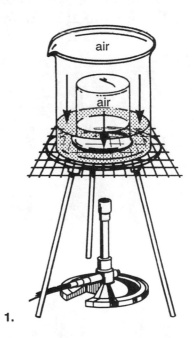

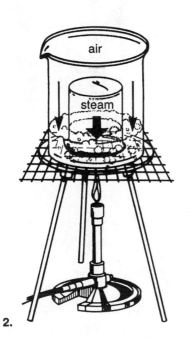

2. As the water boils, bubbles of steam move up into the small beaker, pushing some of the air out. As long as the water is boiling hot, the pressure of the steam in the small beaker prevents water from moving up into the small beaker. Discuss this effect of the boiling water with students, and have them write the appropriate expla-nation under 2 on the student page.

3. Once the beakers begin to cool, the steam pressure in the small beaker drops to zero, and since most of the air in the small beaker was forced out by the steam, you have a much lower air pressure inside the small beaker than you do outside the small beaker. This greater outside air pressure pushes the water out of the large beaker and up into the small beaker. Discuss this result with students, and have them write the explanation under 3 on the student page.

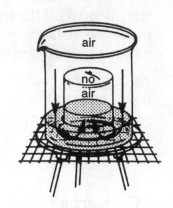

Application

Challenge students to think of a way to cause the water to move up into the small beaker faster. (One way to speed up this process is to stop boiling the water and then immediately put an ice cube on top of the small beaker. By super cooling the small beaker, you speed up the movement of water into the small beaker.)

Take Home

The threat of burns from boiling water and the lack of necessary materials pose problems that are too great for students to attempt this activity outside the classroom.

16 The Problem of the Sunken Egg

Problem: *Can you get the egg to rise again?*

Predict

1. The king's servant is sunk! While bringing the king his morning egg, the servant notices that the egg has sunk to the bottom of the dish. No one may touch the king's food. What is the servant to do? In the space below, list as many ways as you can think of to bring the egg up to the surface of the water. Remember, there is only one rule: You cannot touch the egg directly with your fingers.

Conclude

2. What did the teacher do to solve the problem, and how does the teacher's method work?

---

16 For the Teacher

Objective

In this activity, students will use their powers of observation, critical thinking, and problem solving to meet the challenge of floating the sunken egg.

Materials Needed

- a 1000-ml beaker
- 1 fresh egg
- 1 container of table salt
  (The 26-oz. [737-gm] size will be sufficient.)
- enough water to fill the beaker about 3/4 full
- 1 stirring rod

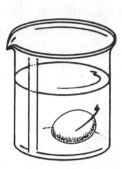

Curiosity Hook

Have the beaker of water with the sunken egg in it set up where students can see it as they come into the classroom.

Setup

1. Fill the beaker about 3/4 full of water. The exact amount of water is not critical.
2. Carefully place a fresh egg in the water. The egg should sink to the bottom of the beaker.
3. Challenge students to think of as many ways as possible to bring the egg back up to the surface of the water. The only rule is that they may not touch the egg directly with their fingers. Have students write their solutions to the problem under 1 on the student page.

Safety Concerns

Use common sense.

Outcomes and Explanations

1. Ask students to share their solutions to the problem with the rest of the class. Accept all reasonable solutions. Try some of these student solutions as time and practicality permit.

2. My solution to the problem is to change the density of the water. The egg sinks because it is just a little denser than water, so if you make the water more dense, you can literally squeeze the egg back up to the surface.

3. To change the density of the water, start adding salt slowly and stirring the water with a stirring rod. Be careful not to break the egg.

4. As the salt molecules dissolve and move into the spaces between the water molecules, the density (compactness) of the water becomes greater than that of the egg, and the egg is forced to the surface. To float the egg, you will need to add about 1/2 or more of the container of salt to the beaker of water.

5. Discuss the differences in density with students, and have them write the explanation under 2 on the student page.

Application

Have students consider the following questions:

1. In some places like the Great Salt Lake and the Dead Sea, it is almost impossible to drown. Why? (These bodies of water have so much salt in them that the density of the water is much higher than that of the human body. Consequently, you would float like a cork.)

2. An ocean-going cargo ship has sailed into the Great Lakes. Would this ship float higher or lower in the Great Lakes than it did in the ocean and why? (The ship would float lower in fresh water than in salt water because salt water is denser than fresh water.)

Take Home

Encourage students to take this activity from the classroom and challenge their parents, siblings, and friends to solve the problem of the sunken egg.

17   The Hanging Hammer

Problem: *The teacher has a hammer that seems to defy gravity. How can this be?*

Observe

1. Carefully observe and, in the space below, accurately describe the hanging-hammer system you see before you.

Predict

2. Is the hammer actually defying gravity? In the space below, list all the possible explanations you can come up with to explain how the hanging-hammer system works.

Conclude

3. Why doesn't the hammer fall? How does this system remain balanced? Explain.

17 For the Teacher

Objective

In this activity, students will use their powers of observation and critical thinking to discover how an object's center of gravity affects its balance.

Materials Needed

- 1 hammer with a wooden handle and a metal head
  (In order for this balancing trick to work, the hammer handle must be made of wood and the head of metal. A rubber mallet or a hammer with a metal handle will not work.)

- 1 wooden ruler or stick of the same approximate size as a ruler
  (Because they bend too much, a plastic ruler will not work.)

- about a 1-foot piece of string or very light wire

Curiosity Hook

Have the hanging-hammer system set up where students can see it as they come into the classroom.

Setup

1. Make a loop in the string or light wire about 4" (10 cm) in diameter.

2. Slip the loop of string or wire around the ruler and the hammer handle.

3. Let the end of the hammer handle press against the ruler. If you have problems with the string or wire slipping up the hammer handle, file a small notch in the handle.

4. Hang the entire system from the edge of a table. The hammer, seemingly in defiance of gravity, should hang and not fall.

ruler or stick

5. Ask students to carefully observe and accurately describe this hanging-hammer system that seems to defy gravity. Have students put their descriptions under 1 on the student page.

6. Challenge students to explain why the hammer does not fall. Have students write their hypotheses under 2 on the student page.

Safety Concerns

Be careful not to drop the hammer on your toes.

Outcomes and Explanations

1. Why doesn't the hammer fall? The answer lies with the principle physicists call *center of gravity*. The center of gravity of any object (or in this case, system) is that point at which all of its weight can be considered concentrated. Regardless of how unbalanced a system looks, if the center of gravity is under the pivot point (the point of support for the system), it is a stable system. If the center of gravity is above the pivot point, the system is unstable. In our hanging-hammer system, the ruler and the hammer handle add little weight, so the heavy metal hammer head moves the center of gravity under the table and below the pivot point.

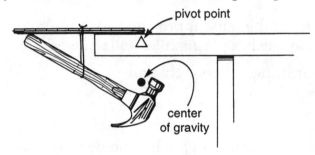

Discuss the principle of center of gravity with students, and have them write the explanation under 3 on the student page.

Application

1. The lower the center of gravity a system has, the more stable the system is. Share this information with students, and have them use it to explain why race cars are designed low to the ground and why tight-rope walkers use long, heavy poles during their routines. (Both the low-to-the-ground design of race cars and the weight and length of the poles of tight-rope walkers lower their centers of gravity, making them both more stable.)

2. To further illustrate the idea that center of gravity, not appearance, determines the stability (balance) of a system, consider building the device illustrated here. You can adjust the position of the two rubber stoppers and the position of the ball on the wire until the device balances. Use a narrow base to rest the device on.

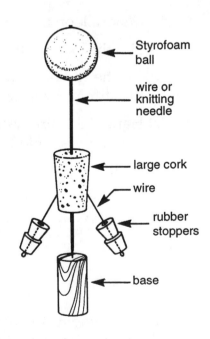

Take Home

If students have access to the necessary materials, encourage them to build the balance system shown in the Application section.

18 Worth Its Weight

Problem: *Is this crown just fool's gold?*

The king has had made a crown of pure gold. However, he suspects that his crown maker has cheated him. The king believes that the crown maker substituted another metal (like copper) for some of the gold and kept the gold for himself.

Is the crown pure gold? The king has summoned you and asked for your help in solving this problem.

Predict

1. Consider the crown made by your teacher. (We will use a clay crown to serve as a model of the real crown.) How could you determine if this model crown was made only of pure gold (or, in this case, pure clay)? In the space below, explain how you would solve the king's problem. You can do anything you want to the crown except cut it up or damage it in any way.

18 For the Teacher

Objective

In this activity, students will use their powers of critical thinking and problem solving in an attempt to solve the problem of the king's crown.

Materials Needed

- modeling clay
- some lead shot or BBs
- a balance or scale
- a large graduated cylinder

Curiosity Hook

Wear a paper crown when students enter the classroom.

Setup

1. Do the following ahead of time:

    a. Make two balls of modeling clay. The exact size of each ball is not critical.

    b. Form a crown from one ball of clay. The crown must be small enough to fit inside the graduated cylinder.

    c. Insert several lead shot pellets or BBs into the clay crown. Carefully smooth the clay so students cannot see that anything has been added to the clay crown.

    d. Weigh both the clay crown and the ball of clay. Add more clay to the ball of clay until it weighs approximately the same as the clay crown.

2. Discuss the problem with students and show them the clay crown you constructed ahead of time. Do not show students the ball of clay you also prepared ahead of time. Students should understand that the clay crown will serve as a model of the real crown and as an aid to help them visualize the problem.

3. Challenge students to solve the problem of the king's crown. Have students write their solution(s) under 1 on the student page.

Safety Concerns

Use common sense.

Outcomes and Explanations

1. This will not be an easy problem for students to solve. A great deal of the success of

this activity will depend on your ability to ask guiding questions and drop revealing hints until some student suggests weighing the crown and comparing the weight of the crown to the weight of the same amount of gold (clay) given to the crown maker.

2. Weigh both the clay crown and the remaining ball of clay. They should both weigh the same. Guided questioning on your part should help students realize that weight alone is not an indication of purity.

3. In the third century B.C., the Greek mathematician and physicist Archimedes solved this very problem and discovered a principle that bears his name. According to Archimedes' principle, a body wholly or partially immersed in a fluid displaces a volume of water equal to its own volume. If the king's crown and an amount of gold equal in weight to what was supposed to be in the crown were both immersed in water, they should displace the same amount of water.

4. Fill the graduated cylinder about half full of water. Immerse the clay crown and read the level of the water displaced by the crown.

5. Remove the clay crown from the graduated cylinder and then immerse the ball of clay. Make sure you have exactly the same amount of water in the cylinder as when you immersed the clay crown. Read the level of the water displaced by the ball of clay.

6. If both readings are the same, the ball of clay and the clay crown are pure clay (gold). However, in this case, the readings should be different because of the impurities (lead shot or BBs) you added to the clay crown. Tear the clay crown apart and show students what was added.

7. Discuss with students how Archimedes' principle applies to this problem. Both crowns weigh the same. Since lead is heavier than clay, however, the clay crown with the pellets contains a smaller amount of material and, therefore, displaces less volume than the ball of clay. Have students write the explanation under 1 on the student page.

Application

1. Legend has it that Archimedes hit on the solution to this problem while immersing himself in a bath. This bolt-from-the-blue solution is said to have so excited Archimedes that he ran naked through the streets shouting, "Eureka! Eureka!"("I have discovered it!")

2. Ask students to explain how and why something floats. Also, have them consider how metal ships, which are much heavier than water, can float. Archimedes' principle of buoyancy explains how objects float. Buoyancy is the loss of weight an object seems to undergo when it is placed in a liquid. The buoyant force is the ability of water (and other fluids) to exert an upward force on an object immersed in it. The buoyant force on an object submerged in a fluid is equal to the weight of the fluid displaced by that object. A block of wood floats because its weight pushing down on the water is less than the buoyant force of the water pushing up. A block of steel sinks because its weight is greater than the buoyant force of water. How then does a

---

steel ship float? The large hollowed-out bowl shape of the ship displaces much more water than the solid block of steel. When the ship displaces enough water to equal the weight of the steel, it floats.

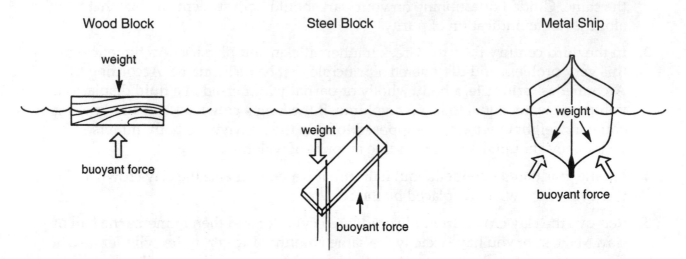

Wood Block   Steel Block   Metal Ship

weight   weight   weight

buoyant force   buoyant force   buoyant force

Take Home

If students have access to the necessary materials, encourage them to take this activity from the classroom and use it to amaze and teach their parents, siblings, and friends.

19 Will the Water Fall Out?

Problem: *Since a window screen has holes in it, water should flow through it easily. Right?*

Observe

1. Carefully observe and, in the space below, accurately describe what happens when water falls onto a piece of window screen.

Predict

2. Will the same thing happen if the screen is placed on top of a jar full of water? The teacher has a canning jar full of water with a piece of window screen for a lid. In the space below, predict what will happen when the teacher inverts the jar.

Conclude

3. What did happen when the jar was inverted, and why did it happen that way?

19 For the Teacher

Objective

In this activity, students will use their powers of observation and problem solving to solve the mystery of why the water will not flow through the holes in a window screen.

Materials Needed

- 1 canning jar with screw ring
  (The exact size of the jar is not critical.)

- 2 pieces of window screen
  (One piece should be about 4 to 6 inches square. The other piece should be cut in a circle to fit the top of the canning jar. Window screen can be purchased at a hardware store or lumberyard.)

- a note card large enough to cover the top of the jar

- enough water to fill the jar

- a washtub or sink to catch water

- several toothpicks

- a faucet or container of water to pour

Curiosity Hook

Run water through the square piece of window screen as students come into the classroom.

Setup

1. Do the following ahead of time:

    a. Cut one piece of screen in a circle that fits the top of the canning jar.

    b. Fill the canning jar completely full of water.

    c. Place the screen circle carefully on top of the jar full of water and screw the ring on securely.

    d. Keep the jar out of sight.

2. Have students observe and describe what happens as water is poured onto the square piece of window screen. Have students write their descriptions under 1 on the student page.

3. Now, show students the jar of water with a piece of screen for a lid. Have students predict what will happen when you invert the jar. Have students write their predictions under 2 on the student page.

4. Place a note card over the top of the jar and hold it tightly in place while inverting the jar. Do this over a washtub or sink.

5. Carefully slide the card off the top of the inverted jar. The water should not come out of the jar through the screen. Have students write this result under 3 on the student page.

6. Poke a toothpick through the screen to show that the screen on the jar does indeed have holes in it. Even poking a toothpick through the screen should not cause the water to come out. However, if you tap on the screen, water will come through the screen. Do this over a washtub or sink.

Safety Concerns

Use common sense; spilled water can make floors slippery.

Outcomes and Explanations

1. Water falling onto the square piece of window screen moves easily through the holes in the screen. So why doesn't the water fall out of jar through the screen when the jar is inverted?

2. The secret lies in the molecules of water. Water molecules have cohesive (attracting) forces and tend to "stick" together. When this happens at the surface of the water, the water molecules become more tightly packed and a thin "skin" forms on the surface. This thin surface layer ("skin") of tightly packed water molecules is called surface tension.

3. The surface tension of the water holds the water molecules tightly enough to prevent them from moving through the holes in the screen when the jar is inverted. Have students write this explanation under 3 on the student page.

4. When you tap on the screen, you disrupt the cohesive forces between the water molecules. This destroys the surface tension effect and water can finally move through the holes in the screen.

Application

Students who have spent time around the edge of a lake or pond may have seen small insects "walking on water" due to surface tension. If you place it carefully, even a steel needle can be made to float on water because of surface tension. Challenge students to investigate on their own the effects that weight and shape have on the ability of objects to float due to surface tension.

Take Home

This activity can be done by students outside the classroom. However, this activity can get messy, so caution them to conduct their experiments in a location where spilling water will not be a problem.

20 The Water Baton

Problem: *Strange things happen when we twirl the water baton. Let's investigate.*

Observe

1. In the space below, describe the water baton.

2. Observe as the teacher puts liquids into the water baton. In the space below, explain what liquids were used and how much of each liquid was placed into the water baton.

Conclude

3. The teacher will now twirl the water baton. What happened to the liquids, and why did this happen?

20 For the Teacher

Objective

In this activity, students will use their powers of observation, critical thinking, and problem solving to solve the mystery of why the bubble appeared.

Materials Needed

- 1 clear glass or rigid plastic tube about 3 feet (1 m) long by about 3/8 inch (10 mm) in diameter
(You may be able to borrow this from the chemistry department of your school.)

- 2 solid rubber stoppers to fit the ends of the tube

- enough distilled water to fill the tube half full
(You may also be able to get this from your chemistry department.)

- enough alcohol to fill the tube half full
(Ethyl alcohol is preferred, but isopropyl [rubbing] alcohol may replace it.)

- food coloring

- optional—sand and enough marbles to fill a small, clear container

Curiosity Hook

Twirl the empty water baton as students come into the classroom.

Setup

1. Show students the empty tube with both stoppers in place. Have them describe this apparatus under 1 on the student page.

2. Remove the stopper from one end of the tube, and fill the tube about half full of colored distilled water.

3. Carefully fill the remaining half of the tube with alcohol, letting the alcohol slowly trickle down the side of the tube. You want as little mixing with the water in the tube as possible. Seal the end with the stopper. There should be no air bubble in the tube when the tube is sealed. If a bubble occurs, remove the stopper and add more alcohol.

4. Show the interface between the two liquids and point out that there is no mixing yet.

5. Now, begin to carefully twirl the tube. Have students note the appearance of a bubble in the tube as the mixing of the liquids occurs. Have students describe this under 3 on the student page.

Safety Concerns

1. Alcohol can cause serious gastric disturbances if taken internally. Dispose of alcohol by pouring it onto a sidewalk and allowing it to evaporate.

2. Twirl the water baton very carefully, and position students a safe distance away as you twirl it.

Outcomes and Explanations

1. Why does a bubble appear? Did some of the liquid leak out? Did some evaporate? No, the answer lies instead with molecular spacing. Molecules have spaces between them. When you mix the molecules of water with the molecules of alcohol, the alcohol molecules slip in between the molecules of water. Thus, the total volume of the mixture becomes less, and a bubble appears.

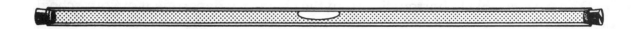

2. Although molecular spaces cannot be seen, this activity proves they do exist. Have students write this explanation under 3 on the student page.

Application

Model this activity for students by pouring marbles (representing the water molecules) into a clear container that is half full of sand (representing the alcohol molecules). The marbles will rest on top of the sand. When the container is full, put a lid on it and turn it over. The sand will filter between the marbles. Students should understand that the marbles and sand only illustrate how molecules of matter behave and that what you are showing them is only a model.

Take Home

Due to the lack of materials and the safety factor of alcohol, it may not be practical to have students conduct this activity outside the classroom.

21 A Dirty Problem

Problem: *Can you think of some new uses for dirt?*

You see it every day. You walk on it because it covers most of the surface of all the land masses on earth. Yet people take it for granted. What is it? Dirt (or more correctly, soil). It's everywhere in large quantities, but is it good for anything?

1. Invent three new uses for dirt (soil). The only rule is, be original! Come up with new ways of using this abundant material that's all around us.

2. Explain your new uses for dirt in the space below.

21 For the Teacher

Objective

Inventors are creative and observant people who see connections, patterns, and solutions that others often overlook. This activity helps develop the critical thinking, problem solving, and "inventioneering" skills of your students by challenging them to dream up new functions for a common substance—dirt (soil).

Materials Needed

- some dirt in a clear container

Curiosity Hook

Pour dirt from your hand into a container as students enter the classroom.

Setup

1. Display a container of dirt where students can see it. This visual aid may help generate ideas and solutions to the problem.

2. Have students brainstorm new uses for dirt, and have them write their solutions to the problem under 2 on the student page.

Safety Concerns

None

Outcomes and Explanations

1. Encourage students to be original and creative in their solutions to the problem. Challenge them to see a common substance in new ways.

2. Encourage each student to come up with at least one new use for dirt. Have students share their new uses for dirt with the rest of the class. List student responses on a chalkboard or large sheet of paper.

Application

Challenge students to test some of the new uses for dirt that class members have come up with. These tests could be performed in the classroom or outside the classroom by students working on their own. You never know. Sitting in your class may be a kid who goes down in history for having the spark of insight that led to a totally new industry revolving around dirt.

22 Blow Up the Balloon

Problem: *Can you meet the challenge and blow up the balloon?*

Observe

1. Carefully observe and, in the space below, accurately describe the balloon-in-a-bottle apparatus the teacher has prepared.

Predict

2. In the space below, list all the ways you can think of to blow up the balloon inside the container. There are two rules:

    a. You may not remove the stopper from the container.

    b. You may not touch the balloon in any way.

22 For the Teacher

Objective

In this activity, students will use their powers of observation, critical thinking, and problem solving to think of ways to blow up the balloon.

Materials Needed

- 1 clear glass or plastic container
  (The exact size of the container is not critical.)

- 1 two-hole rubber stopper that fits the container

- 1 straight piece of glass tubing

- 1 bent piece of glass tubing

- 1 small balloon

Curiosity Hook

Blow up a balloon as students enter the classroom.

Setup

1. Ahead of time, use the materials to set up a balloon-in-a-bottle apparatus like the one illustrated. You may be able to borrow most of the materials needed from the chemistry department of your school.

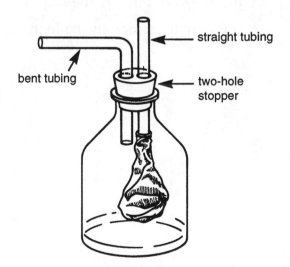

2. Have students observe and describe this balloon-in-the-bottle apparatus. Have students write their descriptions under 1 on the student page.

3. Now, challenge students to think of as many ways as possible to blow up the balloon in the container within the rules set forth on the student page. Have students explain their solutions to the problem under 2 on the student page.

4. Have students share their solutions to the problem with the rest of the class. List student responses on a chalkboard or a large sheet of paper.

Safety Concerns

1. Be careful if you heat and bend glass tubing. Hot glass doesn't look any different than cool glass, but it can cause painful burns.

2. Be careful when inserting the glass tubing into the rubber stopper. Wet the tubing and the stopper for lubrication before insertion. With a cloth, grasp the tubing close to the stopper, and slowly twist the tubing into the stopper. Breaking glass tubing while trying to insert it into a stopper can cause serious lacerations. Check with your chemistry department to see if they have a special safety device that is available for this purpose.

Outcomes and Explanations

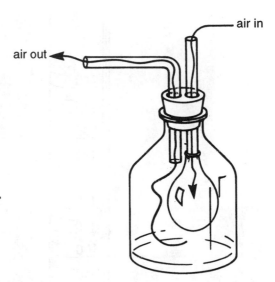

1. The most obvious solution is to blow on the straight piece of tubing. This will indeed inflate the balloon.

2. Less obvious, but just as effective, is to suck on the bent piece of tubing. When you suck air out of the container, you lower the air pressure in the container. The air pressure outside the container is now greater than inside. The greater outside pressure forces air down through the straight tube and into the balloon, causing it to inflate.

Application

Challenge your students further by giving them this problem: Inflate the balloon in the bottle and then keep it from deflating. (One solution is to blow on the straight tube or suck on the bent tube and not remove your lips. Another possibility is to blow on the straight tube or suck on the bent tube then put your finger over the other tube.)

Take Home

It is probably not practical to have students take this activity from the classroom. Students most likely would not have access to the necessary materials, and there is the safety factor of working with glass tubing.

23 A Colorful Column

Problem: *Can you predict the order in which these colorful liquids will layer?*

Predict

1. The teacher will pour the liquids one at a time into the glass column. On the column
    diagrammed below, draw and label the order in which you think the liquids will layer.

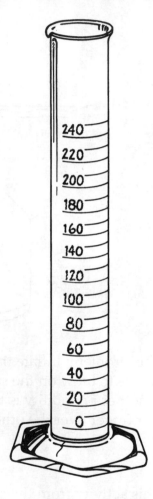

Conclude

2. After seeing this demonstration, would you say there is a difference between density
    (how compact something is) and viscosity (how thick something is)?

---

23 For the Teacher

Objective

In this activity, students will use their powers of observation, prediction, critical thinking, and problem solving to determine the order of layering of liquids in a density column.

Materials Needed

- 1 clear graduated cylinder
  (A 250-ml graduated cylinder would be ideal.)

- a variety of colorful but common liquids:

  maple syrup (brown)

  antifreeze (yellow-green)

  dishwashing detergent (blue)

  shampoo (gold)

  mouthwash (red)

  mineral oil (colorless)

- enough clear containers to hold the colored materials
  (Small beakers work well for this.)

Curiosity Hook

As students enter the classroom, open and arrange the various liquids and small beakers.

Setup

1. Once students are settled and carefully observing, pour each material from its original container into a separate small beaker. This allows the students to note the viscosity (thickness) of each material.

2. Challenge students to predict the order in which the materials will layer when they are poured into the cylinder. Have students draw their predictions on the column illustrated under 1 on the student page.

3. Once students have written their predictions, pour the liquids one after the other slowly down the side of the cylinder. Each material will layer according to its own appropriate density level. Antifreeze is fluorescent, so if you have access to an ultraviolet light, add the antifreeze last and trace its flow with the UV light.

Safety Concerns

1. Antifreeze can cause serious gastric disturbances if taken internally.

2. Dispose of the materials in the density column by flushing them down a sink with a lot of water.

Outcomes and Explanations

1. Without knowing the exact liquids you will be using, there is no way of knowing ahead of time what the exact order of layering will be. Practice ahead of time so that you know what the exact layers will be.

2. Students usually predict that the more viscous materials will be in the lower layers. However, this is usually not what happens.

3. Once this activity has been completed, it should be clear to students that the most viscous material is not necessarily the most dense material. Have students answer question 2 on the student page accordingly.

Application

Challenge students to further investigate the relationship between viscosity (thickness) and density (compactness) by making density columns using other common household materials.

Take Home

Encourage students to take this activity from the classroom and explore the density of a variety of different liquids. How many different layers can they create in one container? Caution students not to experiment with any type of toxic material such as cleaning fluids. Also, if they use antifreeze at home, it is extremely toxic to animals and should not be left where a pet can drink it.

24   Fire and Ice

Problem: *What happens to the temperature of ice water as you begin to heat it?*

Observe

1. What is the temperature of the ice water?

Predict

2. In the space below, predict what will happen to the temperature of the ice water when we start heating it. For our purposes, ice water is defined as water that has ice floating in it.

Conclude

3. Did the temperature of the ice water change when we started heating it? Why did it happen that way?

---

24 For the Teacher

Objective

In this activity, students will use their powers of observation and critical thinking to understand why the temperature of ice water does not rise even while the ice water is being heated.

Materials Needed

- 1 large glass beaker
  (Anything from a 600-ml to a 1000-ml beaker will work nicely.)

- 1 chemical thermometer
  (glass only, not aluminum-encased)

- several trays of ice cubes

- a source of heat
  (If you have access to natural gas outlets, use a bunsen burner. Otherwise, a butane burner or a hot plate will suffice. In a pinch, an alcohol burner will work. However, it will take a long time to heat the water using an alcohol burner.)

- a stand to hold the beaker while heating

- safety goggles

Curiosity Hook

Set up the apparatus and stir the ice water with the thermometer as students enter the classroom.

Setup

1. Shortly before students come in, fill the beaker about 1/2 to 3/4 full of water and add 6–8 ice cubes to the water. Place the beaker on the stand, and position the burner beneath the beaker. If you are using a hot plate as a heat source, set the beaker on the hot plate, but do not turn it on yet.

2. Stir the ice water carefully with the thermometer. Try and time it so that the temperature of the ice water reaches 32° F (0° C) about the time students enter the classroom. Continue stirring as students enter.

3. Once students are settled and ready to begin, read the temperature of the ice water. Have a student record this temperature on a chalkboard or large sheet of paper and label this temperature as the starting temperature. Have each student also record this temperature under 1 on the student page.

4. Now, ask students to predict under 2 on the student page what will happen to the temperature of the ice water when heat is added.

5. Start heating the beaker and continue to stir the ice water with the thermometer. Take a temperature reading every minute. Record these readings minute by minute under the starting temperature recorded on the chalkboard or large sheet of paper.

6. The temperature of the water will not increase until all the ice is gone. Have students write this result as it becomes plain under 3 on the student page.

Safety Concerns

1. You should wear safety glasses while stirring and heating the ice water.

2. Position students a safe distance from the heat source.

Outcomes and Explanations

1. Most students will predict, and logically so, that the temperature of the ice water will increase as heat is added. However, this is not what happens. The temperature of the water will not increase an iota until all the ice has melted.

2. Why? Where did the heat go? The answer lies in what physicists call "phase changes." Water undergoes two phase changes: from ice to water and from boiling water to steam. All the heat energy applied to the water pushes it to the next phase change. Therefore, as you heat ice water, all the heat energy is going into melting the ice. Once the last vestige of ice has melted, the phase has changed, and the heat energy being applied begins to heat the water, driving it to the next phase change— boiling water to steam. Have students write the explanation under 3 on the student page.

Application

Challenge students to consider the following problems:

1. You live on a large lake because you love to fish. You especially love fishing for fork-tailed lake smack. However, these fish only bite as the lake water begins to warm after the winter freeze. When would you fish for this particular species and why? (Phase change is phase change, and it doesn't matter if it happens in a beaker or a lake. The lake will not begin to heat until all the ice has melted. Once that has happened, the lake water will begin to warm, and that would be the time to fish for fork-tailed lake smack.)

2. In this demonstration, we added heat to cold. What will happen if you add ice water (32° F or 0° C) to boiling water (212° F or 100° C)? (Physicists call this the Law of Heat Exchange and it can be expressed as

$$\text{heat lost} = \text{heat gained}$$

The final temperature of the mixture lies between the original temperatures of the ice water and boiling water.)

Take Home

The safety factor of heating water and the necessity of having specialized equipment make it impractical for students to duplicate this demonstration outside the classroom.

25 Sticky Water

Problem: *Has your teacher discovered sticky water that can make a dish defy gravity?*

Observe

1. Carefully observe and, in the space below, accurately describe the apparatus your teacher has set up.

Conclude

2. Explain why the dish of water doesn't fall.

25 For the Teacher

Objective

In this activity, students will use their powers of observation, critical thinking, and problem solving in an attempt to understand why the dish doesn't fall.

Materials Needed

- 1 round dish
  (One half of a plastic petri dish works the best. You may be able to borrow a plastic petri dish from the biology department of your school.)

- enough water to fill the dish

- a smooth surface to "stick" the dish onto
  (A piece of clear glass or plastic about 8 to 10 inches square would be ideal.)

- boxes or books to serve as a support for the piece of glass or plastic

Curiosity Hook

Have the apparatus set up where students can see it as they enter the classroom.

Setup

1. Do the following shortly before students come in:

   a. Place the clear glass or plastic piece on a support of boxes or books.

   b. Fill the dish to overflowing with water.

   c. Carefully bring the dish of water up to the glass or plastic piece and push it against the smooth surface. Make sure you push hard enough to force some of the water out of the dish. There should be no air bubbles in the dish.

   d. Carefully remove your fingers from the dish. The dish should remain attached to the plastic. If it falls, you didn't get a good seal and some air leaked in.

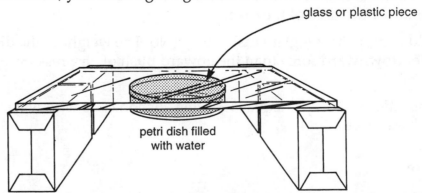

glass or plastic piece

petri dish filled
with water

Once students are settled, have them observe and describe the sticky dish of water. Have students write their descriptions under 1 on the student page.

Safety Concerns

You may want to conduct this demonstration over a pan in case the dish falls.

Outcomes and Explanations

1. Some students may guess that you have glued or somehow attached the dish to the plastic. And, in a sense, that is what has happened.

2. By pushing the dish against the plastic, you removed all the air in the dish. Therefore, the only force working down on the dish is gravity (the weight of the water and the weight of the dish). Pushing up on the dish is the greater force of air pressure.

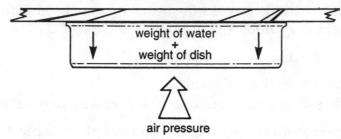

3. Discuss this result with students and have them write the explanation under 2 on the student page.

Application

Challenge students to consider the following problems:

1. Calculate how much air pressure is pushing up on the dish. To do this you have to find the total surface area of the dish.

    a. The bottom of the dish is a circle, so use the formula for finding the area of a circle.

    $$\text{Area} = (3.14) \times (\text{radius of the circle})^2$$

    b. The side of the dish is a long, curved rectangle so use the formula for finding the area of a rectangle.

    $$\text{Area} = \text{height} \times \text{length}$$

    c. Add the area of the bottom to the area of the side of the dish. This will be the total surface area of the dish.

2. Would a large, heavy glass dish work? (No. The weight of the dish would be a greater downward force than the upward push of air pressure, and the dish would fall.)

One application of this principle is wetting suction cups with water to make them stick to smooth surfaces.

Take Home

If students have access to the necessary materials, encourage them to take this activity from the classroom and use it to amaze and teach their parents, siblings, and friends.

26 Is It There?

Problem: *Is air really there? Can you prove it?*

Stop and think for a minute about what you know about air and the atmosphere. We take air for granted, and we assume it's there because that's what we've always been told. But can you prove that air is there? Logic would tell us that if something has weight, it must exist. Therefore, if you prove that air has weight, you prove that air is really there.

Predict

1. Invent a device to show that air has weight. The only rule is to keep it as simple as possible. Once you think you have a solution, use the space below to diagram and explain your device.

2. Once you have your device diagrammed and explained, show it to your teacher for evaluation.

Conclude

3. Describe what the teacher did to prove that air has weight.

26 | For the Teacher

Objective

In this activity, students will use their past experiences and previous knowledge as well as their powers of critical thinking, problem solving, and inventioneering in an attempt to prove that air is really there.

Materials Needed

- 1 balloon

- a scale

- This is a very open-ended activity. As such, there is no way of knowing which directions student creativity and inventiveness will take. Therefore, you will not know what materials students need until after their devices have been invented.

Curiosity Hook

Blow soap bubbles as students enter the classroom. This is a good visual effect and can then lead into questions about what the bubbles are filled with and floating through.

Setup

1. Use the visual effect of the soap bubbles to lead into a brief discussion of air and the atmosphere. Tailor the depth of this discussion to the age and knowledge level of your students.

2. Now, challenge students to invent a device to weigh air. Encourage students to be creative, but urge them to keep their devices as simple and workable as possible. An elaborate design might technically work but may not be practical to construct in class.

3. Once students have a solution to the problem and have diagrammed and explained their devices under 1 on the student page, you should evaluate their designs for feasibility and safety.

4. One of the simplest approaches that you can demonstrate for students is to use a simple scale such as a kitchen scale to weigh an empty balloon. Then blow the balloon up, and weigh the balloon when it is full of air. The inflated balloon should weigh more. Have students describe your method under 3 on the student page.

5. Conclude by constructing and testing the student devices. There are several ways to approach this. If each student designs a feasible device, you could have each student build their own device. If you deem only some of the designs workable or practical, you could have groups of students build those devices.

Safety Concerns

You should act as safety and technical advisor. As students ask you to evaluate the plans for their devices (technical advisor), keep safety uppermost in mind (safety advisor).

Outcomes and Explanations

1. The balloon and scale demonstration weighed a balloon when it was empty and when it was inflated. Since air is the only thing that was added to the balloon, the increase in weight would indicate that the air that was added has weight.

2. There is no way to predict what solutions students will develop to this problem, so you are somewhat on your own in evaluating student designs. Not to worry. Most designs will be easy to evaluate. They clearly will or will not work. Others, however, will appear to be possibly feasible, but the only way to tell will be to actually construct and test the devices. Encourage your students to come up with many creative solutions to this challenge.

Application

Can you think of other ways to prove the presence of air? For instance, if you push a bottle underwater and let it fill up, you can observe bubbles of air rushing out.

Take Home

Encourage students to challenge their parents, siblings, and friends with this problem.

27 Pick It Up

Problem: *Can you pick up the bottle?*

Predict

1. In this activity, the challenge is to pick up the empty bottle you see before you. But where's the challenge in that? You merely grab the bottle and pick it up. Right? Perhaps you should read the rules first. There are two rules:

   a. You cannot touch the outside of the bottle directly.

   b. You only get two kinds of equipment to work with, an empty bottle and several soda straws.

2. Can you pick up the bottle? In the space below, explain as many ways as you can think of to meet the challenge.

27 For the Teacher

Objective

In this activity, students will use their powers of critical thinking, problem solving, and "inventioneering" in an attempt to meet the challenge and pick up the bottle.

Materials Needed

- at least one empty glass soda bottle

- soda straws, flexible or plain

If you have enough empty soda bottles available, give each student a bottle and several straws, and challenge them to work individually on the problem. Otherwise, conduct this activity as a group demonstration.

Curiosity Hook

Have one or two empty bottles with soda straws taped to them at strange angles sitting where students can see them as they enter the classroom.

Setup

1. Discuss the challenge and the rules with students. These are detailed on the student page.

2. Let students brainstorm and work on this challenge for a reasonable length of time. Have students write their solution(s) to the problem under 2 on the student page.

3. Have students share their solutions with the rest of the class.

4. Actually test at least some of the student solutions to the problem as time and practicality permit.

Safety Concerns

Use common sense handling the glass bottles.

Outcomes and Explanations

1. Many student solutions will fail because the soda straws seem too weak to support the bottle.

2. However, soda straws are surprisingly strong if you know their secret, and that secret is that soda straws are strong only if a push or pull is applied along the length of the straw.

---

3. Use that secret by bending the end of a straw, forming a section wider than the mouth of the bottle. Now, slip the bent end into the bottle and maneuver it so the end is wedged against the inside of the bottle.

Using the strength of the straw to withstand a push (on the shorter section inside the bottle) and a pull (on the longer section outside the bottle), you can lift the bottle by the straw with no trouble at all.

Application

Demonstrate the great hidden strength of a soda straw by driving one through a potato. (Hold a potato in one hand and a soda straw, not the flexible type, in the other. With a swift thrust, plunge the straw straight into the potato. With a little practice, you can drive a straw completely through the potato without bending or damaging the straw in any way.)

Take Home

Encourage students to try this activity at home and experiment with different types of containers.

28 Toad in a Hole

Problem: *Can you save the toad?*

Oh no! An extremely rare and endangered species of toad has wandered into a construction site and fallen down a hole. The hole is big enough to stick a hand and arm into, but it's more than three feet deep, so you can't reach your hand all the way down to the toad. You don't want to use a long stick for fear of hurting the toad. What should you do?

Predict

1. In the space below, explain how you would rescue the toad in the hole.

28 For the Teacher

Objective

In this activity, students will use their powers of critical thinking and problem solving in an attempt to rescue the toad in a hole.

Materials Needed

None

Curiosity Hook

If you have access to a live animal, like a toad, an earthworm, or an insect, place it at the bottom of a graduated cylinder or long jar, and place the jar where students can see it as they enter the classroom.

Setup

1. Discuss the problem with students. The problem is detailed on the student page.

2. Let students brainstorm and work on this problem for a reasonable length of time. Have students write their solution(s) to the problem on the student page.

3. Have students share their solutions with the rest of the class.

Safety Concerns

None

Outcomes and Explanations

One possible solution is to slowly drop sand into the hole. The toad will keep moving its feet to stay on top of the sand until the pile gets high enough that the toad can be reached.

Application

Small particles fall into the spaces below large particles, pushing the large particles up. This same principle explains why the whole potato chips rise to the top of the bag, and only the broken ones lie at the bottom.

Take Home

Encourage students to take this activity from the classroom and use it to challenge their parents, siblings, and friends.

29 Cans of Pop Puzzle

Problem: *Can you solve the cans of pop puzzle?*

Observe

1. Carefully observe and, in the space below, accurately describe the situation you see before you.

Predict

2. What is going on? Why does one can of pop float while the other sinks? In the space below, list all the possible explanations you can think of to explain what you see.

Conclude

3. Why did it happen this way?

29 For the Teacher

Objective

In this activity, students will use their powers of observation, prediction, critical thinking, and problem solving in an attempt to solve the cans of pop puzzle.

Materials Needed

- two cans of soda pop of the same type—one regular and one diet
  (Use cans of soda and not plastic containers of soda. We want one container of soda to sink and one to float, but with plastic containers both will float. Furthermore, I have not tested all types of canned soda, but I do know that Coke and Diet Coke will produce the effect we seek.)

- a large clear container
  (I use a five-gallon aquarium.)

Curiosity Hook

Have the system set up where students can see it as they enter the classroom.

Setup

1. Do the following shortly before students enter the classroom:

    a. Fill the container nearly full of water.

    b. Place both cans of soda into the container. The can of regular soda should sink while the can of diet soda should float.

    c. Position this system where students can see it as they enter the classroom.

2. Once students are settled, have them observe and describe the discrepancy between the cans of soda. Have students write their descriptions under 1 on the student page.

3. Now, challenge students to offer possible explanations for the discrepancy between the cans of soda. Have students write their solutions to this puzzle under 2 on the student page.

4. Have students share their explanations with the rest of the class.

Safety Concerns

Use common sense.

Outcomes and Explanations

1. The discrepancy between the two cans of soda can be explained by differences in density.

2. The regular soda has ingredients in it (mainly heavy corn syrups as sweeteners) that the diet soda does not have.

3. This makes the can of regular soda denser than the water and the diet soda, so it sinks. The diet soda, without the heavy syrups, is slightly less dense than the water, so it floats.

4. Discuss this discrepancy with students and have them write the explanation under 3 on the student page.

Application

The health implications ("empty calories") of drinking large amounts of regular soda can be visualized by students if you will open the cans of soda you used in this activity, pour them into separate containers, and allow them to evaporate. The container holding the regular soda will have a thick, dark goo in the bottom while the container holding the diet soda will have practically nothing in it.

Take Home

Encourage students to take this activity from the classroom and use it to amaze and teach their parents, siblings, and friends.

30 Momentum Magic

Problem: *How good are you at seeing patterns?*

Predict

1. Carefully observe the system of hanging balls; in technical terms, this system is called a ballistic pendulum. Predict what will happen when one ball is pulled back and released to hit the other balls.

Conclude

2. What did happen when one ball hit the other balls?

Predict

3. In the space below, predict what will happen when two balls from the same side are pulled back and released and hit the other balls.

Conclude

4. What did happen when two balls hit the other balls?

Predict

5. Predict what will happen when three balls from the same side are pulled back and released and hit the remaining balls.

Conclude

6. What did happen when three balls hit the remaining balls?

Predict

7. Predict what will happen when a ball from each end is pulled back and both are released at the same time, hitting the other balls.

Conclude

8. What did happen when a ball from each side hit the other balls?

9. In the space below, write a scientific "law" to explain what you have just observed. Use the word "momentum" somewhere in your explanation.

30 For the Teacher

Objective

Scientists look for patterns and rhythms in nature in order to unravel the secrets of the universe. In this activity, students will practice and develop their powers of observation, prediction, pattern recognition, and critical thinking as they attempt to verbalize the action and reaction of balls in a ballistic pendulum.

Materials Needed

- 1 ballistic pendulum
  (A commercially-prepared pendulum works best. You could purchase one [they are somewhat expensive] or perhaps you could borrow one. The physics department of your school might have one, or your students might know who has one. If necessary, construct your own.)

Construction Tips

1. It doesn't matter what you use for balls, but the balls must all be identical in size and weight. Commercial pendulums use steel ball bearings, but tennis balls, beach balls, or even basketballs will work. The larger the balls, the easier it will be for students to observe the patterns of collision.

2. The balls should be aligned as straight as possible from top to bottom and from side to side.

3. Use a total of five balls. This will create a challenging problem later in the activity.

Curiosity Hook

Swing the balls on the pendulum as students enter the classroom.

Setup

1. Pull back one ball from one end. Ask students to predict what will happen to the other balls when you release the ball you are holding. Have students write their predictions under 1 on the student pages.

2. Release the ball. When it hits the other balls, a single ball will fly out from the other side—one ball in, one ball out. Have students write this result under 2 on the student pages.

3. Now, pull back two balls from one end. Ask students to predict what will happen to the other balls when you release the two balls you are holding. Have students write their predictions under 3 on the student pages.

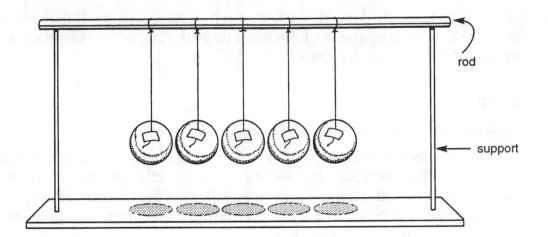

4. Release the two balls. When they hit the other balls, two balls will fly out from the other side. Have students write this result under 4 on the student pages.

5. Pull back three balls from one end. Ask students to predict what will happen to the remaining two balls when you release the three balls you are holding. By now, students think they see the pattern, so this setup presents a challenging problem as only two balls remain to be hit. Have students write their predictions under 5 on the student pages.

6. When three balls are released and hit the remaining two balls, three balls will fly out from the other side (the one in the middle switches)—three balls in, three balls out. Have students write this result under 6 on the student pages.

7. Now, really complicate things by pulling back a single ball from each side. Ask students to predict what will happen to the remaining balls when you release one ball from each side at the same time. Have students write their predictions under 7 on the student pages.

8. The pattern will continue. When one ball is released from each end, one ball will fly out from each end—one ball in, one ball out. Have students write the result under 8 on the student pages.

9. Challenge students to write a scientific "law" based on the pattern they have seen that explains the behavior of the balls. They should include the word *momentum* somewhere in their explanation.

Safety Concerns

Use common sense when releasing the balls.

Outcomes and Explanations

1. This series of events demonstrates the law of conservation of momentum. Simply put, this law states that in collisions such as you observed with the balls, the total momentum of the colliding bodies is not changed; x amount of energy put into the system yields the same x amount of energy coming out of the system.

2. Discuss with your students the "laws" that they have written. If they did not come up with a similar explanation, have them write the law of conservation of momentum under 9 on the student pages.

Application

1. Challenge students to consider the following question:

   What if the objects colliding together and moving apart are of different sizes? (Momentum is still conserved. Suppose you have two balls colliding, and one ball is twice as big as the other ball. The smaller ball will bounce away at twice the speed of the larger ball, which will move away in the opposite direction. Application of this can be seen in the results of head-on collisions between trucks and cars, where the truck driver almost always survives the collision, but the car passengers do not survive.)

2. An important application of the principle of conservation of momentum can be seen in the launching of a rocket. When a rocket fires, hot exhaust gases are expelled through the rocket nozzle. Momentum equal in magnitude is imparted to move the rocket in the opposite direction.

Take Home

If students have access to the materials necessary to construct their own pendulum, encourage them to take this activity from the classroom and try it on their own.

31 Hot Air Balloon

Problem: *What will happen to the balloon when we turn up then turn off the heat?*

Observe

1. Carefully observe and, in the space below, accurately describe the hot air balloon system the teacher has prepared.

Predict

2. In the space below, predict what will happen to the balloon when it is placed over boiling water.

Conclude

3. What did happen to the balloon when it was placed over boiling water, and why did it happen that way?

Predict

4. In the space below, predict what will happen to the balloon when the water begins to cool.

Conclude

5. What did happen to the balloon as the water cooled, and why did it happen that way?

31 For the Teacher

Objective

In this activity, students will use their powers of observation, prediction, critical thinking, and problem solving to understand why the balloon ends up inside the flask.

Materials Needed

- a 150–200 ml Erlenmeyer flask
  (The chemistry department of your school should have this item of glassware.)
- several large balloons
- a hot plate or burner and stand
- a heat-resistant pad or plate
- beaker tongs or hot pads
- safety goggles

Curiosity Hook

Blow up and release balloons as students enter the classroom.

Setup

1. Pour about 20 ml of water into the flask. Attach the balloon to the flask and set the flask on the hot plate or stand.

2. Ask students to observe and accurately describe this system. Have students write their descriptions under 1 on the student page.

3. Now, ask students to predict what will happen to the balloon when the water is heated. Have students write their predictions under 2 on the student page.

4. Remove the balloon, turn on the hot plate or burner, and begin to heat the water. Let the water boil vigorously for at least one minute.

5. Turn off the heat and use beaker tongs or hot pads to move the flask to a heat-resistant pad or plate.

6. Quickly place the balloon over the mouth of the flask. This can be a bit tricky as the flask and the steam coming out of the flask will be very hot, but it is imperative that the opening of the balloon is lined up with the opening of the flask. If not, the balloon may be pinched off, and the demo will not work.

7. The balloon should begin to expand. Have students write this result under 3 on the student page.

8. Once the balloon is fully inflated, ask students to predict what will happen to the balloon as the water begins to cool. Have students write their predictions under 4 on the student page.

9. As the water begins to cool, the balloon will deflate and even be pushed down into the flask. Have students write this result under 5 on the student page.

Safety Concerns

1. You should wear safety goggles while heating the water.

2. Position students a safe distance from the heat source and the hot flask.

3. Use caution when placing the balloon on the flask of boiling water. The flask and the steam coming from the flask will be very hot.

Outcomes and Explanations

1. When water is heated it changes state from a liquid to a gas. As it changes to this state, it expands. This expansion of liquid water into steam is what inflates the balloon. Discuss the effect heating the water has on the balloon, and have students write this explanation under 3 on the student page.

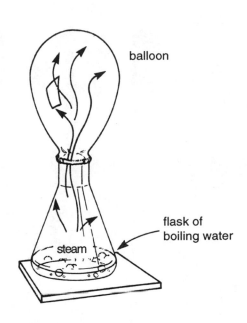

balloon

flask of boiling water

steam

2. As the water boils and steam pours out the opening of the flask, the steam carries some of the air in the flask with it. This reduces the amount of air in the flask.

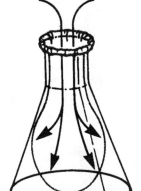

greater air pressure

3. As the water begins to cool, the steam pressure is lost, and since you now have less air inside the flask than outside the flask, the balloon is pushed down into the flask by the greater outside air pressure. Discuss this result with students and have them write the explanation under 5 on the student page.

Application

Challenge students to consider the following questions:

1. How can we get the balloon back out of the flask without touching the balloon? (If you begin to heat the water again, steam pressure will force the balloon out of the flask).

2. How does steam pressure affect our daily lives? (Steam pressure is used to turn turbines that generate electricity; some of your students may live in homes or buildings that are heated by steam pressure.)

Take Home

Lack of materials and the safety factor of boiling water make it inadvisable to have students take this activity from the classroom.

32 The Mystery of the Clay Sticks

Problem: *Why does one clay stick float while the other sinks?*

Observe

1. Carefully observe and, in the space below, accurately describe the situation you see before you.

Predict

2. List all the possible explanations you can think of as to why one clay stick floats while the other sinks.

Conclude

3. Why does one clay stick float while the other sinks? What is the explanation?

32 For the Teacher

Objective

In this activity, students will use their powers of observation, prediction, and critical thinking in an attempt to discover why one clay stick floats while the other sinks.

Materials Needed

- some modeling clay
- a short wooden stick or dowel
- 2 clear glass or plastic containers
- enough water to fill the containers about 3/4 full
- a balance or scales
- optional—a ball of clay about 1 inch in diameter

Curiosity Hook

Have the setup where students can see it as they come into the classroom.

Setup

1. Do the following ahead of time:

   a. Apply just enough clay over the stick or dowel to hide the wood but not so much that it becomes heavy enough to sink. This is clay stick A.

   b. Carefully weigh the clay-covered stick. Then weigh out an equal weight of clay only. This will be clay stick B.

   c. Form clay stick B into the same approximate shape and length as clay stick A.

   d. Fill each container about 3/4 full of water.

   e. Place each clay stick in a separate container. Clay stick A (clay + wood) should float while clay stick B (clay only) should sink.

   f. Position the containers where students can see them as they enter the classroom.

2. Have students observe and describe this setup. Have students write their descriptions under 1 on the student page.

3. Challenge students to explain why one clay stick floats while the other sinks. Inform them that both clay sticks have the same mass (weight). Have students write possible explanations under 2 on the student page.

Safety Concerns

Use common sense.

Outcomes and Explanations

1. Write the following formula on a chalkboard or large sheet of paper:

$$\text{Density} = \frac{\text{mass (weight)}}{\text{volume (space occupied)}}$$

2. Water is considered to have a density of 1.0 grams per cubic centimeter. Objects float or sink because they are less or more dense than water.

3. Both clay sticks have the same approximate mass (weight), but the density of the floating clay stick is smaller than the one that sinks because the hidden wooden stick or dowel increases the volume.

4. Challenge students to do the math:

    Imagine that clay stick A has a weight of 20 grams and a volume of 22 cubic centimeters. What is the density of clay stick A? (0.91 gm/cc)

    Imagine that clay stick B has a weight of 20 gm and a volume of 10 cc. What is the density of clay stick B? (2.0 gm/cc)

    Why does A float and B sink? If water has a density of 1.0 gm/cc, A is less dense than water, so it floats; but B is more dense than water, so it sinks.

5. Discuss the differences in density with students, and have them write the explanation under 3 on the student page.

Application

Show students a ball of clay about 1 inch in diameter. Challenge them to come up with a way to make the clay float without changing the amount of clay. (Form the clay into a boat shape. When placed in water, it should float because you increase the volume [space occupied] when you change the shape.)

Take Home

If students have access to modeling clay, encourage them to take this activity from the classroom and use it to amaze and teach their parents, siblings, and friends.

33 The Sagging Cord Puzzle

Problem: *Can you pull the cord perfectly straight?*

Predict

1. The teacher has tied a strong cord around a heavy book. If two students lift the book by pulling on the ends of the cord, will they be able to make the cord perfectly straight? Write your prediction in the space below.

Conclude

2. What happened when the students pulled on the ends of the cord, and why did it happen that way?

33 For the Teacher

Objective

In this activity, students will use their powers of observation, prediction, and critical thinking in an attempt to understand why the cord always sags.

Materials Needed

- a heavy book
- about 6 feet of strong cord

Curiosity Hook

Twirl a piece of cord as students enter the classroom.

Setup

1. Tie the middle of the cord around the book.

2. Select two students to grasp the ends of the cord. Ask the rest of the class to predict whether or not the selected students can pull the cord straight. Have students write their predictions under 1 on the student page.

3. Now, have students tug on the cord and try to pull it perfectly straight. No matter how hard they pull, the book will always make the cord sag slightly in the middle. Let several pairs of students try this impossible feat. Have students write this result under 2 on the student page.

Safety Concerns

Use common sense.

Outcomes and Explanations

1. The pull on each end of the cord has two components, an upward (vertical) pull and an outward (horizontal) pull.

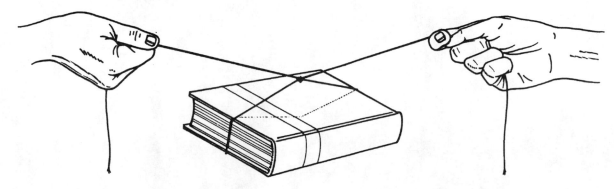

2. The outward (horizontal) pulls on each end of the cord balance the book horizontally. It is the vertical pull that lifts the book. However, as the book is lifted and the cord becomes more horizontal, the vertical pull lessens. The force of gravity pulling down on the book is stronger than the vertical pull. Therefore, it becomes impossibly difficult to pull hard enough to lift the book the last bit needed to make the cord exactly straight. In a tug-of-war between muscle power and gravity, gravity always wins (eventually). Discuss this concept with students and have them write the explanation under 2 on the student page.

Application

In the above activity, will the size of the book or the size of the cord make a difference in the outcome?

Take Home

Encourage students to take this activity from the classroom and use it to amaze and teach their parents, siblings, and friends.

34 A Messy Mixture

Problem: *Can you separate the mixture back into its original parts?*

Oh no! Your little cousin, Trudy Tumtater, has made a mess. While you were supposed to be watching her, she mixed table salt, iron filings, sawdust, and sand together.

Predict

1. Can you think of a way to separate and recover each of the original four items? Explain your plan below.

34   For the Teacher

Objective

In this activity, students will use their powers of critical thinking and problem solving, plus draw on their personal experiences, to solve the messy-mixture separation challenge.

Materials Needed

- a small quantity of each of the following will be needed:

  sand

  table salt

  iron filings

  sawdust

- a magnet

- a filter or wire mesh colander

- several small beakers or clear containers

- one large beaker or clear container

Curiosity Hook

Place a small quantity of each of the above items in separate beakers or clear containers. Have the beakers or containers where students can see them as they come into the classroom.

Setup

1. Use the scenario about the messy mixture given on the student page or make up one of your own. As you are describing the scenario, pour the sand, table salt, iron filings, and sawdust from their beakers or clear containers into a single larger beaker or clear container.

2. As you shake or stir these components together, challenge students to devise a way to separate this mixture and retrieve the original components. On the student page, have students write and/or diagram their plans to meet this challenge.

Safety Concerns

Use common sense.

Outcomes and Explanations

1. Here is one possible solution. First, use a magnet to separate the iron filings from the mixture.

2. Then add water to the remaining three items. The sawdust will float and can be skimmed off. The salt will dissolve in the water.

3. Filter the water containing the sand and the dissolved salt. The filter traps the sand.

4. The water containing the dissolved salt can be heated (or left standing for several days) until the water evaporates, and the salt can be recovered.

Application

This might be an appropriate time to discuss the differences between an element, a mixture, and a solution. An element is made of all the same kind of atoms. For example, a bar of pure gold would be an element because it contains only gold atoms. There are 92 naturally occurring elements. Materials made from combining the atoms of two or more elements are called mixtures. One common mixture is concrete. Concrete is a mixture of cement, gravel and/or sand, and water. Students may be wearing another example. Many modern fabrics are a mixture of polyester and cotton. Liquid mixtures are called solutions. White vinegar is a solution containing acetic acid and water. Have students come up with other examples of elements, mixtures, and solutions that are a part of our everyday lives.

Take Home

Students could certainly challenge their parents, siblings, and/or friends with this activity. If students wish to actually do the separation at home, they will need access to the four items that make up the mixture—sand, table salt, iron filings, and sawdust—and the other equipment.

35   A Sound Reproducer

Problem: *Can you make the record play?*

Imagine that the security of this country and life as we know it depends on you quickly getting sound off an old record. You have an old record turntable that works, but the arm holding the needle is gone. Can you solve the problem and get sound off the record?

Predict

1.   In the space below, explain your method for solving this problem.

Conclude

2.   What did the teacher do to solve the problem, and how does the teacher's method work?

35   For the Teacher

Objective

In this activity, students will use their powers of critical thinking and problem solving in an attempt to get sound from an old record.

Materials Needed

- an old phonograph record
  (Do not use a valuable old record as it can be damaged by the pin.)
- a record turntable
- a paper or Styrofoam cup
- a pin or needle

Curiosity Hook

Have the record turning on the turntable when students enter the classroom.

Setup

1. As the record turns on the turntable, present the scenario and the problem outlined at the start of the student page.

2. Challenge students to get sound from the record. Have them write their solutions to the problem under 1 on the student page.

3. Have students share their solutions with the rest of the class.

4. Push a pin or needle into the center of the bottom of a paper or Styrofoam cup.

5. Now, lightly hold the cup in your hand and touch the pin to the record as it turns. The paper cup will produce sound just like a loudspeaker. Have students write this result under 2 on the student page.

Safety Concerns

Use common sense.

Outcomes and Explanations

1. How can a paper cup with a pin in it produce sound? There are grooves in the record. The grooves are not straight and smooth, but ripple from side to side. The ripples are too small to be seen with the unaided eye. When the point of the pin or needle rides in the groove, it vibrates from side to side as it follows the ripples.

2. The vibrations are picked up by the cup and transmitted to the air, which carries

them to your ear. The sound is loud enough to be heard because the surface of the cup is large enough to move a lot of air. Have students write the explanation under 2 on the student page.

Application

1. The large sounding board behind the strings in a piano works the same way as the cup. The board vibrates when a string is struck by the piano key. Its broad surface moves so much air that the sound is made loud.

2. Challenge students to make the sound coming from the record even louder. (This can be accomplished by shaping a large piece of paper into a cone, folding the end, and forcing a needle or pin into this fold. When the needle in this cone is held lightly on the record, the sound should be greatly amplified. Historically speaking, such horns were the forerunner of our modern speakers.)

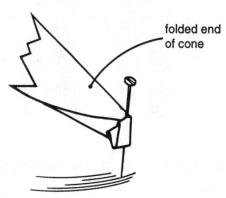

folded end of cone

Take Home

If students have access to the necessary materials, encourage them to take this activity from the classroom and use it to amaze and teach their parents, siblings, and friends.

36 Which One Is the Magnet?

Problem: *Can you identify which object is the magnet?*

Predict

1. Carefully observe the two identical bars the teacher is holding. One of these bars is a magnet and the other is not. Without using any other materials, how can we tell which one is the magnet? Write your solution to this problem in the space below.

Conclude

2. How did the teacher solve the problem?

36 For the Teacher

Objective

In this activity, students will use their powers of critical thinking and problem solving, and their past experiences with magnets, in an attempt to identify the magnet.

Materials Needed

- a strong bar magnet
- an unmagnetized steel bar identical to the bar magnet

Curiosity Hook

Use a magnet to pick up paper clips or some other small metal objects as students enter the classroom.

Setup

1. Show the students the identical bars.

2. Using only the bars and no other materials, challenge students to find out which bar is the magnet.

3. Have students write their solutions to the problem under 1 on the student page.

4. Have students share their solutions with the rest of the class.

5. Demonstrate the alignments shown in the Outcomes and Explanations.

Safety Concerns

Use common sense.

Outcomes and Explanations

1. This alignment proves nothing.

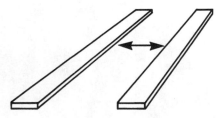

2. This alignment also proves nothing.

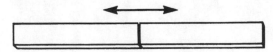

3. The only alignment that can solve
   the problem is this one.

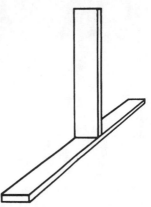

4. In the first two alignments there is no repulsion, so there is no way to tell which bar
   is the magnet. However, in the third alignment, two things can happen which allow
   us to deduce which bar is the magnet.

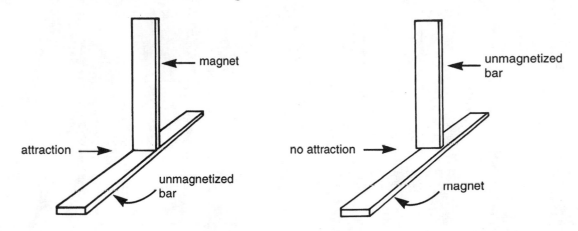

If the magnet is placed in the middle of the unmagnetized bar, the bar is attracted to
the magnet. If the unmagnetized bar is placed in the middle of the magnet, there is
no attraction to the steel in the magnet. Discuss this difference with students and
have them write the explanation under 2 on the student page.

Application

Challenge students to investigate the role magnets play in their everyday lives; for
example, magnets are used to seal refrigerator doors.

Take Home

If students have access to the necessary materials, encourage them to take this activity
from the classroom and use it to amaze and teach their parents, siblings, and friends.

Creative Challenges

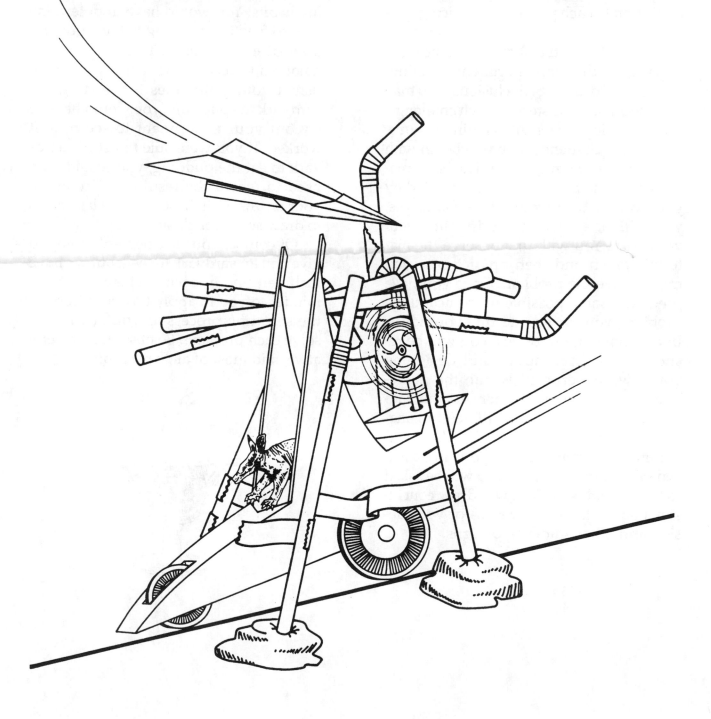

Creative Challenges Introduction

In each Creative Challenge, the teacher will present a problem (challenge) for you to solve. The teacher will give you a set of guidelines to follow for each challenge. Working within those guidelines, it will be up to you to meet the challenge and solve the problem. As you design and develop a solution to each problem, practice applying the science process skills demonstrated by the teacher in the Dynamo Demos: observing, predicting, experimenting, eliminating, and drawing conclusions. You may not use all of these steps in each challenge, but the basic process will remain the same.

In each challenge, you will begin with a specific problem to solve. The basic materials for the project will be provided by your teacher, but in some of the challenges, you will be allowed to add materials of your own. You will first observe the materials given and then predict how you might solve the problem. Once you have predicted some possible outcomes to the problem, you can start testing your ideas. In the challenge activities, you will be responsible for setting up and designing your own experiment within the guidelines given by your teacher. As you construct your design, you will need to analyze the different variables involved and eliminate alternatives until you have a reasonable solution. Finally, you will be asked to test your design. You will draw conclusions about the effectiveness of your design and other students' designs.

Suppose you had to solve the following problem: Do aardvarks eat chocolate pudding? If the teacher gave each student an aardvark and chocolate pudding, you would each need to set up an experiment to solve the problem. You would have to observe the aardvark to determine his eating habits. You would have to determine how to feed the pudding to the aardvark, how often to feed it to him, in what amounts to feed it, etc. Once you established your guidelines for feeding the aardvark the pudding, you would have to try out your experiment to see how it worked. If you were able to get the aardvark to eat the pudding, you would then need to record the results and compare your results with those of your classmates to draw some conclusions.

Of course, your teacher isn't going to give you an aardvark to feed, but we have included one in each challenge as a reminder to you to apply the basic steps of the scientific process as you attempt to solve each problem. Be observant, be creative, and most of all, have fun!

---

1 Wacky Wire

Challenge

Invent a new use for a wire coat hanger.

Rules

1. One wire coat hanger will be supplied by your teacher. You must supply everything else needed to build your invention.

2. Your invention must serve some purpose or have some useful function other than hanging clothes.

3. The coat hanger may be bent, cut, or modified in any way you want. You may use the entire coat hanger or only a portion of it.

4. No more than four other objects or kinds of materials (including glue, tape, etc.) may be used in combination with the coat hanger. The coat hanger may also be used by itself.

5. Brainstorm your design then make a detailed labeled drawing of your invention. Think SAFETY.

6. Once you have your invention designed and drawn, explain it to your teacher. If your teacher decides that your design is safe and practical, you may go ahead and begin building.

7. Once each of you has completed your invention, you will explain your invention to the rest of the class and demonstrate how it works.

1 For theTeacher

The Challenge

Inventors are creative, observant people who see connections, patterns, and solutions that others often overlook. This activity fans the spark of invention in your students by challenging them to dream up a new function for a common object—a coat hanger.

The Rules

1. Have students work alone or in teams.

2. You will need to supply one wire coat hanger per student or per team.

3. Depending on your time restrictions, let students either work on this activity during class time or have them brainstorm, design, and build their inventions outside of class but present and demonstrate them to the rest of the students during class time.

4. After students have had time to brainstorm and come up with a design for some sort of invention, require them to submit a labeled diagram and explanation of their invention before you give them clearance to begin construction. Think safety and practicality.

5. Once construction is complete, have students explain their inventions to the rest of the class using their labeled diagrams. Then, if practical, have them demonstrate how their inventions actually work.

Safety Tips

Discourage designs that have sharp, pointed projections, and avoid those that employ flying projectiles.

Stumped?

If you have a student(s) who gets stuck for an invention idea, offer the following slightly silly suggestions to get the creative process going:

• a giant paper clip

• a back-scratcher

• a small lightning rod

• a roasting stick for cocktail weiners

Awards and Recognition

Display students' design diagrams and, if practical, actual inventions where students, teachers, and parents can appreciate them.

   © 1996 Critical Thinking Books & Software • P.O. Box 448, Pacific Grove, CA 93950 • 800-458-4849

2 The Leaning Tower

Challenge

Design and build the tallest freestanding structure possible.

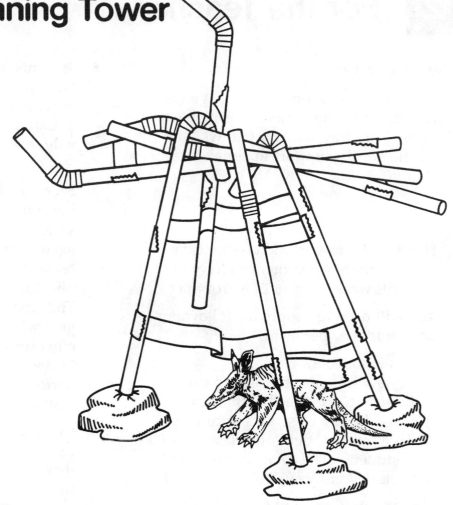

Rules

1. The teacher will provide the following construction supplies:

   • as many plastic soda straws as needed

   • a measured amount of modeling clay

   • 12 inches of masking tape

2. The structure must be freestanding. That is, it cannot be held up by any outside means nor can it be attached to or leaning against anything else.

3. The structure must support only its own weight.

4. The structure must stand long enough to be measured.

5. Structures will be measured from the floor to their highest point. If a structure stands but leans over, the designer(s) can straighten the structure for measurement.

6. Structures must be built within the time limit set by the teacher.

7. The tallest structure will be declared the winning design.

2 For the Teacher

The Challenge

In this activity, students are challenged to build a freestanding structure using a lot of flimsy building material (straws) and very little sturdy support materials (clay and tape).

The Rules

1. Have students work alone or in teams. I find it more convenient to have my students work in teams of two or three.

2. You will need to supply the following per student or per team:

    - plastic soda straws
      (I allow my design teams to use all the straws they need within reason. Three boxes of straws should be more than enough for a class of 12–14 students. The cooks in our school cafeteria kindly supply my straws.)

    - a measured amount of clay
      (I give each team 25 grams of clay. Less clay would make this activity even more challenging.)

    - 12 inches of masking tape

3. I have had students construct structures over 9 feet tall, so plan to work in an area with adequate ceiling height.

4. Clay can make a mess and be somewhat difficult to clean off the floor. Consider putting down newspapers first and then have your students construct their structures on them.

5. You will need a measuring tape. A 12-foot tape should suffice, but I prefer a 20-foot tape just in case.

6. Students may need something to stand on in order to add to their structures as they grow taller. Provide as many chairs and stools as practical. I also provide several 8-foot stepladders, courtesy of our school custodial staff.

7. Set a time limit for construction. I give my students one class period to construct their structure. I issue the challenge and let the students pick their teams; I also prepare the building supplies for each team, except for the tape. The next day as teams come in, they get their piece of tape, pick up their other supplies, and begin construction. Leave enough time at the end of the period for final measurements and cleanup.

8. I allow my teams to measure whenever they want. After any measurement, they can continue adding to their structure. This will help prevent having to measure all the structures at once in a rush at the end of the period.

9. Measure structures from the floor to the highest point. I do not require that the structures stand upright, only that they stand. If a structure is standing but bent over, I allow the team to straighten the structure for measuring. I do not, however, measure structures that bend over so far that they touch and are supported by the floor.

Safety Tips

Urge caution when students are standing on chairs or climbing up and down stepladders. In their haste, someone may fall.

3 Build a Barge

Challenge

Design and build a device that will float as many pennies as possible.

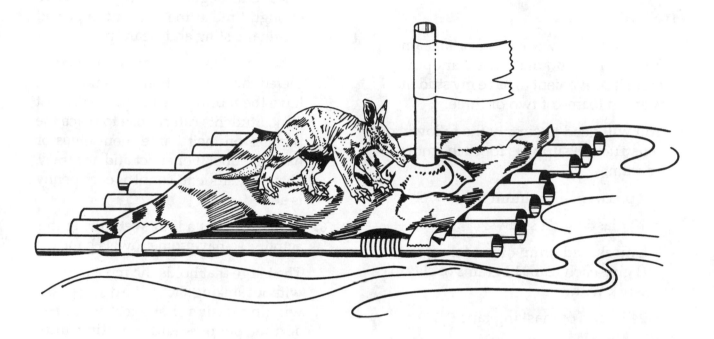

Rules

1. Your teacher will give you the following construction supplies:

    • 2 pieces of aluminum foil, each 6" square

    • 10 plastic soda straws

    • 1 measured lump of modeling clay

    • 24 inches of masking tape

2. You may use all of or part of the construction supplies as needed for your design.

3. Your barge must be constructed to fit within the test tank.

4. Your barge must be completed within the time limit set by the teacher.

5. Each barge will be floated in a test tank. Pennies will be placed one at a time on each barge until the barge sinks.

6. The barge that holds the most pennies without sinking will be declared the winning design.

3 | For the Teacher

The Challenge

In this nautical activity, students are challenged to build a flotation device using minimal materials.

The Rules

1. Have students work alone or in teams. From a time and materials standpoint, I find it convenient to have my students work in teams of two or three.

2. You will need to supply the following construction supplies for each student or team:

    • 2 pieces of aluminum foil 6" × 6"

    • 10 plastic soda straws

    • 1 measured lump of modeling clay (I give each team 10 grams of clay to work with.)

    • 24 inches of masking tape

3. You will also need to set up a tank of water and get some pennies ahead of time. A 3- or 5-gallon aquarium should suffice, but a 10-gallon aquarium would be large enough to accommodate any extra-large barges. The record to date for my students is 149 pennies, so 200 pennies should be more than adequate.

4. Set a time limit for construction. I give my students one class period to construct their barges. I issue the challenge and let the students pick their teams; I also prepare the building supplies for each team, except for the tape. The next day as the teams come in, they get their piece of tape, pick up their other supplies, and begin construction. Leave enough time at the end of the period for final testing and cleanup.

5. When students or teams have completed the construction of their barges, have them bring the barges to the test tank. Students can choose to place the pennies on their barge themselves or have me do it. For effect and accuracy, I have my students count as each penny is added.

6. Things can get a little wet during this activity, so have some towels handy.

7. The barge that holds the most pennies without sinking should be declared the winning design. I suggest you keep adding pennies and counting until each barge totally sinks. Occasionally, barges will have water come over the top but not totally sink until a few more pennies are added. Students should be made aware in advance how you will handle this situation should it arise.

Awards and Recognition

Display a chart showing the number of pennies each student's or team's barge floated.

Bombs Away!

Challenge

Design and construct a device that will catch a fresh egg without breaking it when the egg is dropped.

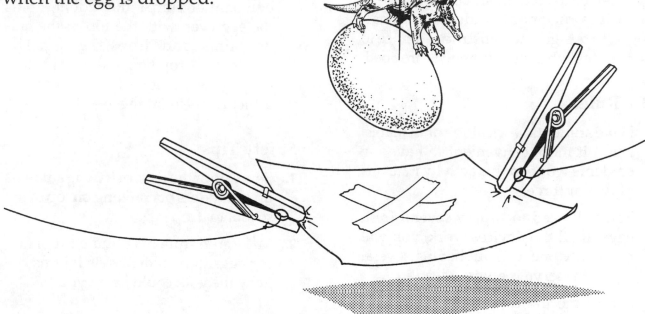

Rules

1. The teacher will supply only the egg(s) to be dropped. You must supply all other construction materials. There is no restriction on what materials you can use.

2. Your device can be no more than 1-foot square and no more than 6 inches tall.

3. A fresh egg with nothing attached will be dropped onto your device from a height determined by the teacher. Your device must be designed to "catch" and prevent breakage of the egg. Your egg faces three dangers:

    *   **poor aim**
        If you miss your egg catcher when you drop your egg, the egg is doomed. There is nothing you can do

    about this danger in your design, but practice with an egg-sized object might help.

    *   **impact**
        Even if you hit your egg catcher with the egg, the impact may break the egg. Remember, it's not the fall that breaks the egg, it's the sudden stop. Design accordingly.

    *   **bounce**
        Even if you hit your egg catcher and the impact doesn't break the egg, the egg may bounce off, hit the ground, and break. Again, design accordingly.

4. Your egg catcher must be built within the time limit set by the teacher.

4 For the Teacher

The Challenge

Egg-drop activities are standard inventioneering challenges. This one, however, presents a twist. Instead of designing a device to contain and protect an egg when it is dropped, students are challenged to design and construct a device to catch an egg when only the egg is dropped.

The Rules

1. Have students work alone or in teams. I find it more convenient to have my students work on this activity in teams of two or three.

2. You will need to supply several fresh eggs. If all the devices work, you will need only one egg, but having several eggs in reserve is a wise idea.

3. Set a time limit. I have my students work on their designs on their own outside of class for several days to perhaps a week. They bring their devices to school on the day scheduled for the activity.

4. I suggest you do this activity outside on an asphalt or concrete surface.

5. I require one member of each team (designated as the bombardier) to drop the egg from the top of a 10-foot stepladder. The egg catcher is placed on the ground next to the stepladder. The bombardier climbs the ladder, holds the egg even with the top of the ladder, aims, and drops the egg. The greater the drop height the more challenging the activity is, but also the more difficult it is to hit the target.

Safety Tips

1. Be aware of the risks of having students up on ladders or crawling on top of the school to drop eggs.

2. Make sure students stand clear of falling eggs. If a student gets injured, legally the yolk could be on you.

Awards and Recognition

1. If practical, display student or team egg catchers where they may be appreciated by other students, teachers, and parents.

2. Award appropriate gag gifts, such as chocolate eggs, to all participants.

5 | Airmobile

Challenge

Design and construct a car that will travel as far and/or as fast as possible using only the force of air from a window fan.

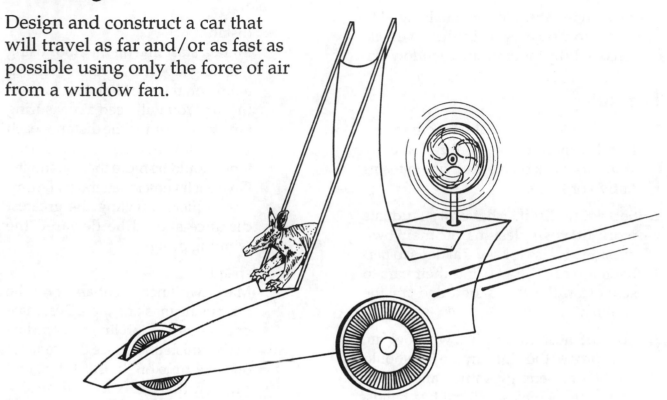

Rules

1. The teacher will supply the window fan, but you must supply all the other supplies needed to construct your airmobile.

2. There are no size restrictions. Your airmobile and its parts can be any size you wish.

3. Your airmobile must have at least 3 wheels.

4. Your airmobile must roll or skid along the floor. It cannot fly through the air.

5. Your airmobile will be placed on the start line and a fan turned on. The airmobile that travels the greatest distance and/or is the fastest, using only the force of air from the fan, will be de-

clared the winning design. You will be given as many trial runs as time permits. Only the longest and/or fastest runs will be counted.

6. You will be allowed to move the fan, but the fan cannot touch the vehicle nor can the fan be moved beyond the start line. The force of air from the fan can be the only source of power used to move your vehicle.

7. Once your car takes off from the start line, you will not be allowed to touch or help your vehicle in any way. No outside assistance or guidance is allowed.

8. Your airmobile must be built within the time limit set by the teacher.

5   For the Teacher

The Challenge

In this activity, students are challenged to use common objects to build a vehicle powered only by the air from a window fan.

The Rules

1. Have students work alone or in teams. I find it more convenient to have my students work on this activity in teams of two or three.

2. Set a time limit. I have my students work on their designs on their own outside of class for several days to perhaps a week. They bring their cars to school on the day scheduled for the activity.

3. You will need to supply a window fan. The force of the air from the fan should be the only energy source allowed to move the vehicles. No rubber bands, electric motors, etc. should be allowed. I do allow students to move or angle the fan at their discretion, but the fan should remain behind the start line at all times.

4. I put no size restrictions on how large the vehicles or the various components of the vehicle can be. However, I do require that each vehicle have at least three wheels.

5. Two categories of competition could be held:

   a. distance
   Mark a start line on the floor using masking tape. Place vehicles one at a time on the start line and turn on the fan. You will need a measuring tape to determine the distance each vehicle travels; a 20-foot measuring tape should be more than adequate. Give each vehicle several trial runs. The vehicle traveling the greatest distance should be declared the winning design.

   b. speed
   Mark two lines 8 feet apart on the floor using masking tape. Designate one line as the start line and one line as the finish line. Place the vehicles one at a time on the start line, and turn on the fan. You will need a stopwatch to determine how long it takes each vehicle to reach the finish line. Give each vehicle several trial runs. The vehicle covering the distance between the start and finish lines the fastest should be declared the winning design.

Safety Tips

In their excitement to move and aim the fan, students could trip over the fan cord.

6 Fantastic Fliers

Challenge

Make a single sheet of paper travel as far as possible.

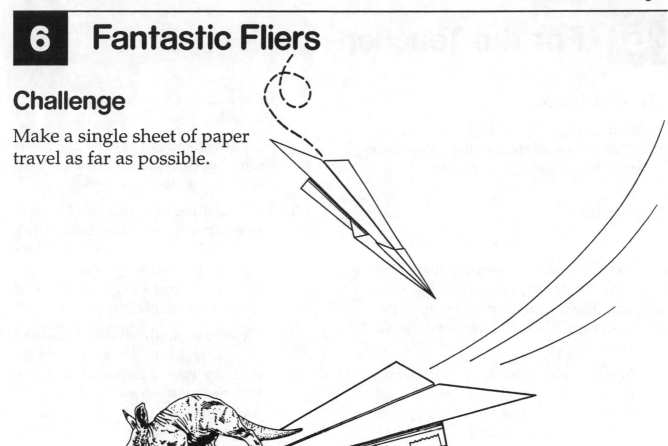

Rules

1. You will be given two sheets of paper, both the same size.

2. Your "airplane" must be made from only one sheet of paper or the pieces cut from one sheet of paper. Use one sheet of paper for practice and the other for the actual competition.

3. The sheet of paper may be folded or cut in any way you choose.

4. Tape, staples, and/or paper clips may be used to hold pieces of your "airplane" together. Nothing else may be attached to your plane. Remember—the heavier the design, the faster it will fall.

5. Paper wads are not allowed.

6. Only one "airplane" per person or group may be entered.

7. Each plane will be hand launched but cannot be touched or aided in any way after launch.

8. Flight distance will be measured from the starting line to where the plane first touches.

9. Each person or group will be allowed three trial runs. Only the longest of the three trial runs will be counted.

10. The plane that travels the greatest distance will be declared the winning design.

11. Your plane must be built within the time limit set by the teacher.

6 For the Teacher

The Challenge

In this activity, you will see the many creative methods students use to make a piece of paper travel as far as possible.

The Rules

1. Have students work alone or in teams.

2. Provide each student or team with two sheets of paper, both the same size. The size of the paper is up to you. Standard 8.5 × 11-inch paper works well.

3. You will also need to provide tape, a stapler, and paper clips in case students or teams wish to incorporate them into their designs.

4. Require student designs to at least somewhat resemble an airplane. Do not allow students to merely wad up the paper and throw it.

5. Set a time limit. I allow my students to work on their designs at school during class time. I set aside several periods for design, construction, and testing and one period for the competition.

6. Hold the competition indoors in a large area like a gymnasium or outside in a parking lot if there is no wind.

7. Each student or someone from each team must hand launch the plane—assisted takeoffs using rubber bands or model rocket engines are not allowed.

8. Use masking tape to mark a start line on the floor. Launch all planes from this line.

9. Give each student or team three trial runs, but count only the longest distance of the three attempts.

10. A measuring tape will be needed to determine flight distance. Our track coaches kindly loan me a 200-foot tape for this activity. Shorter tapes will work, but you may have to move them several times.

11. A variation on this theme is to challenge students to design a plane for accuracy. Place a target a reasonable distance away (as determined by your distance competition) and see which plane can come closest to the target. Use a system of scoring similar to the one below:

on the target	=	10 points
within 6 inches	=	9 points
within 12 inches	=	8 points
within 18 inches	=	7 points
within 24 inches	=	6 points
within 30 inches	=	5 points
within 36 inches	=	4 points
beyond 36 inches	=	no points

Safety Tips

None

Stumped?

If available, books on making paper airplanes might serve as inspiration.

7 Strummin' on the Old Banjo

Challenge

Build a musical instrument using common objects found around the house.

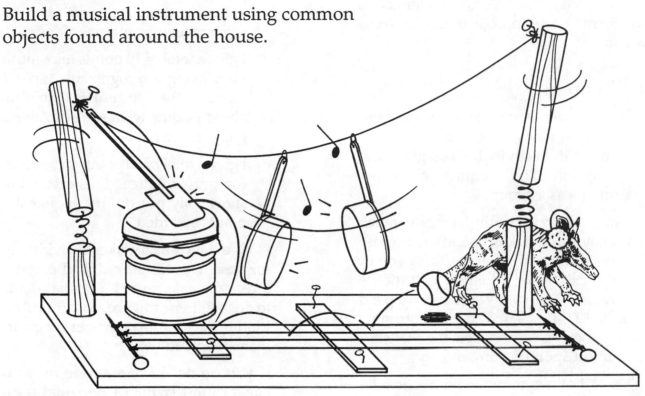

Rules

1. You must supply everything needed to build your music maker. Use common objects found around your house. The teacher will set a limit on how much money can be spent to purchase construction supplies.

2. Commercial musical instruments are not allowed.

3. Your teacher will set a time limit on the design and construction of your music maker.

4. Your music maker will be judged on the following point system:

   design of music maker =
   0–10 points

   number of notes played =
   0–10 points
   (maximum of 10 different notes)

   playing a recognizable melody =
   0–10 points

   performance (put on a show) =
   0–20 points

5. You may perform with other students as a group, but each member of the group will be judged separately on his/her performance.

6. The music maker with the highest total points will be declared the winning design.

7 | For the Teacher

The Challenge

In this activity, students are challenged to transform common objects into musical instruments.

The Rules

1. Have students work alone or in teams. I prefer to have my students work alone on this activity, but I do give them the option of performing with other students as a group.

2. I suggest you set a limit on how much money students can spend to purchase construction supplies for this activity. Your individual situation will dictate this limit. I allow my students to spend only $2.00 on supplies and strongly urge them to try and meet the challenge without spending any money.

3. Do not allow commercial musical instruments. Some student may just go to the mall and buy a plastic flute.

4. Set a time limit on design and construction of the music makers. You can have students bring their construction materials to school and work on their music makers during class, or you can have them work on their own outside of class.

5. I judge the music makers on a maximum 50-point total:

   a. I give a total of 10 points maximum for design. I favor designs that are unique (the wackier the better) and those that reflect the obviously large amounts of thought and effort that went into their construction.

   b. I give a total of 10 points maximum for the number of <u>different</u> notes played up to 10 notes; otherwise, some sharp student will play the same note over and over forever.

   c. I give a total of 10 points maximum for playing a recognizable tune. It helps if the student informs us ahead of time what song he/she is going to try to play.

   d. I give a total of 20 points maximum for performance. The better the show they put on, the higher the points awarded.

   I judge each music maker individually for design, notes played, and playing a recognizable tune. I allow students to do their performance as a group, but I judge each group member's performance separately.

   If you do not have a strong musical background (which I personally do not), you might consider having your music teacher help judge certain aspects of this activity.

6. The music maker with the highest total points should be declared the winning design.

Stumped?

If you have a student(s) who gets stuck for an invention idea, offer the following suggestions to get the creative process going:

- wire harp
- one-string guitar
- water or wood chimes
- hose flute
- finger piano

Racing Blimps

Challenge

Design and construct a balloon device that will fall as fast as possible.

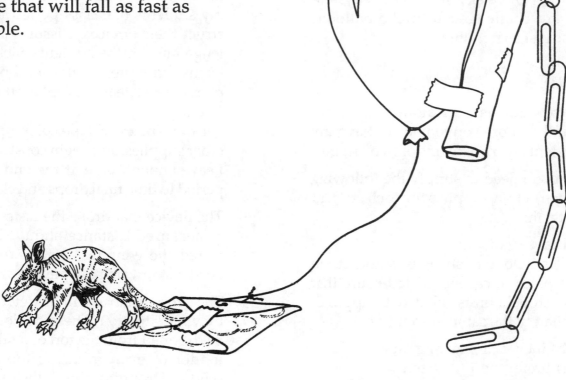

Rules

1. Your teacher will supply the following construction supplies:

   - one balloon
   - 24 inches of masking tape
   - 2 pieces of paper, 8.5 × 11 inches in size
   - 10 paper clips
   - 15 pennies
   - scissors to cut tape and/or paper

2. The scissors cannot be part of your device.

3. All balloons will be inflated with the same amount of air.

4. The construction supplies can be attached to each other or to the balloon in any pattern. You do not have to use all the construction supplies.

5. Your device cannot be designed to pop the balloon on purpose.

6. Your balloon device must be completed within the time limit set by the teacher.

7. The balloon device that falls the fastest from a given distance will be declared the winning design.

8 For the Teacher

The Challenge

We usually picture balloons as floating slowly up to the sky. In this activity, students are challenged to make balloons drop as fast as possible.

The Rules

1. Have students work alone or in teams. I find it more convenient to have my students work in teams of two or three.

2. You will need to supply the following construction supplies for each student or team:

    • 1 large balloon
      (Use a loop of string or a hole cut in a piece of cardboard to ensure that each balloon is filled with approximately the same amount of air.)

    • 24 inches of masking tape
      (I prefer the 1/2" size.)

    • 2 pieces of paper, 8.5" × 11" in size

    • 10 paper clips
      (All clips must be the same size.)

    • 15 pennies

3. The construction supplies may be attached to each other and/or to the balloon in any pattern. All of the construction supplies do not have to be used.

4. Designs that cause the deliberate destruction (popping) of the balloon should not be allowed.

5. Set a time limit for construction. I give my students one class period to construct their devices. I issue the challenge and let the students pick their teams; I also prepare the building supplies for each team, except for the tape. The next day as teams come in, they get their piece of tape, pick up their other supplies, and begin construction. Leave enough time at the end of the period to time final drops and cleanup.

6. The device that drops the fastest from a measured distance should be declared the winning design. You will need a stopwatch to time the drops.

7. The drop distance will be determined by your situation. These devices may be dropped from the top of a tall stepladder or even from the top of the school. Dropping the device a short distance, as by standing on a chair, makes it difficult to accurately time the drop. The important thing is that all balloons be dropped the same distance.

Safety Tips

Be aware of the risks involved with having students clambering up and down stepladders or climbing on top of the school.

9 Crazy Cantilever

Challenge

Design and build an arm of soda straws that will bend the least amount possible when a 100-gram weight is attached.

Rules

1. The teacher will supply the following materials:

   - 20 plastic drinking straws
   - 20 straight pins
   - 24 inches of masking tape
   - scissors as needed to cut straws and/or tape

2. Only the straws, pins, and tape may be part of your arm, but not all of the straws, pins, or tape must be used.

3. The base of your arm can be started on or near the edge of a table or desk and must be taped down to the table or desk. Build the rest of your arm outward from there. A cantilever is a horizontal arm which extends beyond its point of support.

4. The length of the arm and the kind and amount of bracing are up to you. Remember: the longer the arm, the more it will bend when the weight is attached.

5. The following formula will be used to determine which arm holds a 100-gram weight with the least amount of bending:

   length of arm sticking out – sag of arm with weight = rigidness

6. One of your pins must be stuck in the very end of the arm. This is where the 100-gram weight will be attached.

7. The arm with the highest rigidness number will be declared the winning design.

8. Your arm must be completed within the time limit set by your teacher.

9 For the Teacher

The Challenge

In this activity, students are challenged to build a cantilever support with minimal materials while working against the clock.

The Rules

1. Have students work alone or in teams. I find it more convenient to have my students work in teams of two or three.

2. You will need to supply the following for each student or team:

   - 20 plastic drinking straws (The size is up to you, but I find the jumbo size less challenging than smaller sizes.)

   - 20 straight pins

   - 24 inches of masking tape (I recommend the 1/2 inch size.)

   - scissors as needed to cut straws and/or tape

3. Set a time limit. I give my students one class period to construct their arms. I issue the challenge and let the students pick their teams. I prepare the building supplies, except for the tape. The next day as students come in, they get their building supplies and tape and begin construction. Leave enough time at the end of the period for final measurements and cleanup.

4. Have students build their arms on or near the edges of tables or desks. The base of the arm should be taped to the top of the table or desk. Caution students to use a minimal amount of their tape attaching the base; save as much tape as possible for the arm and braces. Let students decide how far back from the edge of the table or desk to start building the base of their arm.

5. You will need a 100-gram weight configured for hanging from the pin in the end of each arm. I have a weight with a wire hook. You could also wrap a weight in tape and attach the tape to the pin in the end of each arm.

6. Students will take two approaches to this problem. Some will go for length at the expense of rigidness while others will build very short heavily-braced arms striving more for rigidness than length. Often, those who go for length will find that their arms totally collapse when the weight is attached.

7. Use the formula on the student page to determine the rigidness of each arm. Use two yardsticks to do this. First, measure the length of the arm projecting from the base with one yardstick before the weight is attached. Next, place one end of the yardstick on the table or desk next to the arm and parallel to it. Now, attach the weight to the end of the arm. Use the other yardstick to measure down from the horizontal yardstick to the end of the arm to determine the amount of bend. This takes several sets of hands, so have students hold the horizontal yardstick while you measure the bend of the arm with the second yardstick. If an arm bends perpendicular, or nearly so, to the horizontal yardstick, it should be regarded as a collapse.

Safety Tips

Urge students to use good safety procedures when working with scissors and straight pins.

Awards and Recognition

1. It really isn't practical to save and display the arms themselves, but do display a chart showing the length, sag, and calculated rigidness of each arm where they may be appreciated by other students, teachers, and parents. Those arms that fail should just be labeled as having collapsed.

10 | Marble Mania

Challenge

Construct a maze that will take a marble from
the top of a shoebox to the bottom of the box
in 30 seconds.

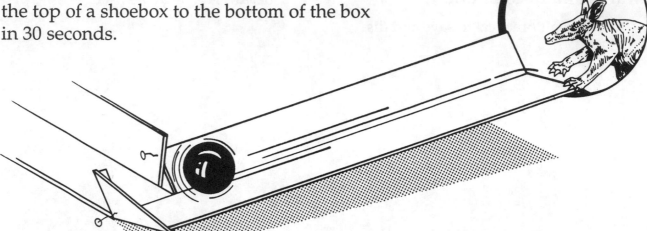

Rules

1. The teacher will provide you with the
   following materials:

   • one shoebox

   • heavy paper and thin cardboard for
     building the maze

   • glue and tape

   • scissors for cutting and shaping
     ramps

   • a marble for practice runs

2. The maze system must start in any top
   corner of the box.

3. A marble-sized entrance hole must be
   made in the top corner of the box directly above where the maze system
   starts.

4. The maze system should be constructed mainly of heavy paper and/
   or thin cardboard. Other items such as
   rubber bands, pins, foam rubber, etc.
   are allowed. You must supply any additional construction supplies needed.
   To be on the safe side, clear your de-

sign ideas with your teacher before you
begin construction.

5. The lid of the box must come off so the
   inside can be examined.

6. A marble-sized exit hole must be made
   in the lower corner of the box where
   the maze system ends.

7. Your maze system must be built within
   the time limit set by the teacher.

8. Time will start when the marble is
   dropped in the entrance hole at the top
   of the box and stop when the marble
   rolls out the exit hole at the bottom.

9. As time permits, you will be given several trial runs. You may make adjust-
   ments to your maze system between
   trial runs. Points will be kept for each
   run. After all runs have been completed, the points for each maze system will be totaled and averaged. The
   maze system that has the highest
   average point total will be declared the
   winning design.

10 For the Teacher

The Challenge

In this activity, students are, in essence, building a timing device using a shoebox and marble.

The Rules

1. Have students work alone or in teams. I find it more convenient to have my students work in teams of two or three.

2. You will need to supply the following for each student or team:

   • one shoebox
   (Ask students to bring these from home or ask local shoe stores to donate some.)

   • paper and cardboard of assorted thickness for constructing ramps

   • white glue and tape

   • scissors

   • a marble for practice runs

   • I do allow other objects such as rubber bands, nails, foam rubber, etc. to be used. However, students must supply any additional construction supplies, and I require them to clear their design ideas with me before they begin construction.

3. Set a time limit. I have my students work on their designs on their own outside of class for several days to perhaps a week. They bring their maze systems to school on the day scheduled for the activity.

4. A hand-held stopwatch will work for

timing. Start timing once the marble is dropped into the entrance hole and stop once it comes out the exit hole.

5. As time permits, give each student or team several trial runs. Allow them to adjust or alter their ramp system between trial runs.

6. Score each run. I use the following point system:

30 seconds	=	10 points
31 or 29 seconds	=	9 points
32 or 28 seconds	=	8 points
33 or 27 seconds	=	7 points
34 or 26 seconds	=	6 points
35 or 25 seconds	=	5 points
36 or 24 seconds	=	4 points
37 or 23 seconds	=	3 points
38 or 22 seconds	=	2 points
39 or 21 seconds	=	1 point

7. Whatever scoring system you use, make the students aware of how points will be awarded before the competition begins.

8. At the end of all trial runs, total each student's or team's points and average. The maze system with the highest point average should be declared the winning design.

Safety Tips

Urge students to practice proper safety procedures when working with scissors.

11 Write It—Do It

Challenge

Turn accurate observations into usable instructions.

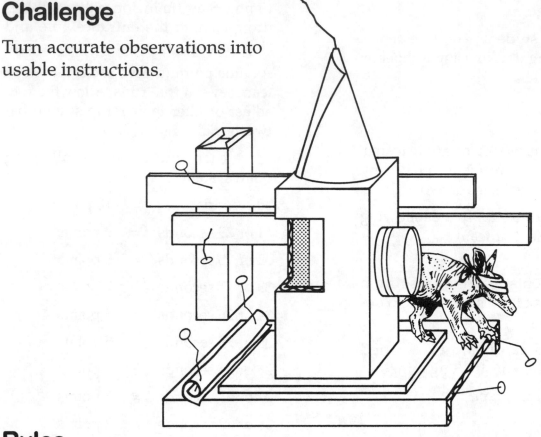

Rules

1. You will work as a team with another student.

2. One team member will be the observer/writer and the other team member will be the doer.

3. The teacher will show an object to the observer/writer only. The observer/writer will be allowed to carefully examine and observe the object.

4. The observer/writer must then write a set of instructions for building an object exactly like the one built by the teacher. No diagrams or pictures are allowed. The object will then be taken apart.

5. The doer must now take the observer/writer's instructions and build an object exactly like the one built by the teacher using the parts of the teacher's original object.

6. A perfect reconstruction of the teacher's original object is worth 50 points. Each mistake made in reconstruction will be a 5-point deduction.

7. The teacher will provide another contraption and you will switch roles.

8. The team with the highest combined point total will be crowned as the champion observers/writers/doers.

11 For the Teacher

The Challenge

Everyone has problems following instructions, but in this activity, students will also come to appreciate how hard it is to write usable instructions.

The Rules

1. Have students work alone in teams of two. Designate one student as the observer/writer and the other student as the doer.

2. Show only the observer/writer an object you have constructed. Tinkertoys or Lego blocks work well for this. The observer/writer should not watch you construct the object but see it for the first time only in its final state.

3. Give the observer/writer adequate time to observe and investigate your object. The observer/writer should then write a set of instructions for building such an object while he/she disassembles your original object.

4. Once the instructions are written and your original object has been disassembled, bring the doer in. The doer must use the instructions and the dis-

assembled pieces to reconstruct your original object.

5. Determine the points earned.

6. Build or provide another object and allow the students to switch roles.

7. Again, determine the points earned and the total. The team with the highest combined point total is the champion observer/writer/doer.

8. A variation of this activity would be to have the observer/writer read the instructions to the doer. Thus, you could challenge students to follow written and/or verbal instructions.

Safety Tips

Use common sense.

Awards and Recognition

1. Display a chart showing the point totals earned by each team.

2. Display some of the contraptions each team was challenged to reconstruct.

3. Display appropriate certificates of achievement made to the winning teams.

12 Bizarre Beans

Challenge

Grow the strangest looking plant possible within a set time.

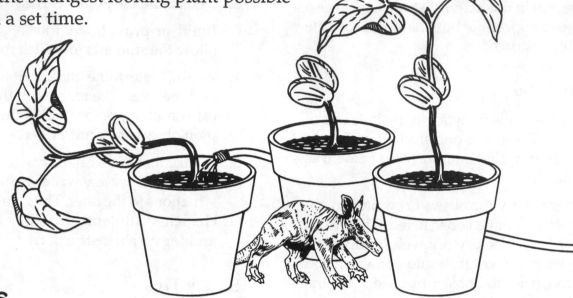

Rules

1. The teacher will give you at least 5 seeds. You must supply everything else.

2. You will need to make the following decisions about your seeds and the plants that develop from these seeds:

   • what material will you plant your seeds in—soil, sand, etc.?

   • what kind of container(s) will you plant your seeds in?

   • how many seeds will you plant in each container?

   • how deep will you plant the seeds?

   • how much light will the plants receive?

   • how much and how often will you water the plants?

   • will you fertilize the plants?

   • if you fertilize the plants, how much and how often?

   • and, most important, how can you regulate and manipulate the plants and/or the conditions the plants are grown in to get the strangest looking plant in the shortest amount of time?

3. Your teacher will set a time limit on growing your plants. Before you begin, decide what you will try to do to your plants to get strange growth patterns. Your teacher will share some information with you that should help you decide what to do to your plants.

4. A word of caution—while plants need some water, and you might decide to try fertilizer, go easy on these things. Your plants will suffer more from too much water and fertilizer than from too little water and fertilizer.

5. The plants will be judged at the end of the time limit set by the teacher.

12 For the Teacher

The Challenge

In this activity, students delve into the biological realm by manipulating the pattern of plant growth to produce the strangest plant possible within a set time limit.

Background Information

Share the following pertinent information with students before they begin this activity: Plants, like all other living creatures, respond to the conditions around them. Such responses in plants are called tropisms. Plants exhibit an amazing number of tropisms. A few plant tropisms are as follows:

- phototropism
  A plant growth response to light coming from one direction. An example is a plant next to a window bending toward the light coming in the window.

- thigmotropism
  A plant growth response to contact with a solid object. An example is the way some vines grow around posts or trees.

- gravitropism (or geotropism)
  A plant growth response to gravity. Roots grow down and stems grow up.

- hydrotropism
  A plant growth response to water. An example occurs when roots encounter water-filled pipes and grow into the pipes rather than around them.

Regulating and controlling these tropisms can produce strange-looking plants.

The Rules

1. Have students work alone or in teams.

2. Provide each student or team with at least 5 seeds. Lima beans, green beans, or pinto beans will work. Peas are also satisfactory, but dwarf types will not produce the longer stems of regular varieties. I prefer mung beans for their rapid growth and elongated stems. Check local suppliers or seed catalogs for mung beans, or order from one of the following sources:

   Connecticut Valley Biological
   82 Valley Road
   P.O. Box 82
   Southampton, MA 01073

   Carolina Biological Supply Company
   2700 York Road
   Burlington, NC 27215

   Ward's Natural Science Establishment, Inc.
   P.O. Box 92912
   Rochester, NY 14692-9012

3. Students should decide which tropism they wish to regulate and control and how they plan to do so.

4. If appropriate and necessary, you may need to supply students or teams with planting medium and/or planting containers. If a student or team plans on using soil as their planting medium, commercial potting soil is best. Soil from a garden or road ditch may contain fungi that can cause rotting seeds or damping off (a fungal disease causing seedlings to wilt and die) of the seedlings.

5. Allow students to place their plant growth containers in the classroom or at home in places which they think will provide maximum growth conditions.

6. You may face the situation where none of a student's or team's seeds germinate,

or if they do, they quickly die. Have a plan to deal with such possibilities if they should arise.

7. Set a reasonable time limit. At least several weeks may be necessary to get satisfactory results. Of course, the time limit can always be altered if necessary. To speed things up, you might plant the seeds yourself, allow the seeds to germinate, and then give your students or teams a container or containers of emerging seedlings.

8. At the end of the time limit, judge the plants. Number all the plants and then have the class vote by secret ballot on which plant they think is the strangest looking. Students or teams should not vote for their own plant. You might also ask other students, fellow staff members, parents, and/or appropriate community patrons to be part of the judging process. The plant receiving the most votes should be declared the strangest-looking plant (*Bizzaro botanico*).

Safety Tips

1. Caution students about putting seeds in their mouths or allowing younger children in the household to do so. Choking could result. Also, some seed companies coat seeds with a fungicide (usually a bright pink color) that could be harmful if ingested.

2. Urge students to handle fertilizers safely. Fertilizer containers should be stored out of the reach of small children and fertilizer/water mixtures should not be left where they might be accidentally ingested by someone.

3. If students build plant growth chambers from flammable materials and place incandescent lights inside, there is a risk of fire.

Stumped?

1. Students might try manipulating phototropism. They might grow plants with the light source from the side or have plants grow through a maze in a shoebox.

2. Students might try manipulating thigmotropism. Plants could be wrapped around wire frames as they grow. We often do this activity around Valentine's Day and some student or team will wrap plants around heart-shaped wire frames. Plants might also be entwined as they grow. A word of caution—rapidly growing stems are quite brittle and break very easily.

3. Students might try manipulating gravitropism (geotropism). The whole pot could be laid on its side or turned upside down, forcing the plants to grow around the pot and then upward.

Awards and Recognition

1. You might consider giving the winning student or team a small houseplant(s) as a prize.

13 Captain Nemo's Contraption

Challenge

Design and build a device that will sink and then resurface.

Rules

1. You must supply everything needed to build your device.

2. The entire device must sink.

3. The entire device or any part of the device must resurface in a reasonable amount of time.

4. Water will be the liquid used to test your device. No harmful or dangerous liquids will be allowed.

5. You may not aid or assist your device in any way once it is placed in the water.

6. Your device must be built within the time limit set by the teacher.

7. Devices that successfully sink and then float will be declared to have met the challenge of Captain Nemo.

13 For the Teacher

The Challenge

In this activity, students actually have two challenges, one easy and one hard. The easy part is getting their device to sink. The hard part is getting all or part of the device to resurface.

Background Information

Captain Nemo was the builder and commander of the submarine *Nautilus* in the classic fiction thriller book *Twenty Thousand Leagues Under the Sea*, written by Jules Verne.

The Rules

1. Have students work alone or in teams.

2. Students should supply all necessary construction supplies for this activity.

3. Set a reasonable time limit for the resurfacing of the devices, and inform students of this time limit before construction begins. If the device has not resurfaced within a few minutes after sinking, it probably never will come up.

4. Set a time limit on construction of the devices. I have my students work on their devices outside of class for sev-

eral days to perhaps a week. They bring their devices to school on the day scheduled for the activity.

5. Consider having each student or team explain to the rest of the class how their device is supposed to work before testing the device.

Safety Tips

Use only water to test the devices. Caustic or dangerous liquids should never be allowed.

Stumped?

The following are a few approaches my students have taken to this challenge:

- Freeze a penny and a small piece of wood in an ice cube. The cube will sink and then melt, and the wood will surface.

- Drill a hole in a seltzer tablet. Attach a weight and a small float (like a fishing bobber). The weight sinks the whole device. The seltzer tablet melts and the float pops to the surface.

- Glue a tiny piece of balsa wood to an aspirin tablet. The tablet sinks and when it dissolves, the wood pops to the surface.

14 | Ping Pong Fling

Challenge

Design and construct a device that will fling a Ping Pong ball as far and/or as accurately as possible.

Rules

1. You may use any or all of the following materials to construct your flinger:

   - 1 Ping Pong ball
   - 1 mouse trap
   - 1 plastic spoon
   - 4 straws
   - 6 popsicle sticks
   - 4 paper clips
   - 6 rubber bands
   - 12 inches of masking tape

   The teacher will explain which, if any, of these materials the teacher will provide and which, if any, of these materials you must provide.

2. Your flinger must be constructed with only the materials given, but you do not have to use all the materials.

3. Your flinger must be completed within the time limit set by the teacher.

4. Your flinger must be resting on the floor when it is fired. You can hold the flinger down (or stabilize it) with your hands and fingers.

5. You will be given as many flings as time permits.

14 For the Teacher

The Challenge

In this activity, students are challenged to use common objects to invent a device that will fling a Ping Pong ball.

The Rules

1. Have students work alone or in teams. I find it more convenient to have my students work on this activity in teams of two or three.

2. You will need to decide which, if any, of the construction supplies you will provide to the students and inform them accordingly.

3. I do not put any size restrictions on the flingers. Students do not have enough construction materials to build anything of any great size.

4. Set a time limit. You might discuss the challenge, brainstorm, and begin construction at one class meeting and then finish construction and conduct the competition at the next class meeting.

5. The flinger should be resting on the floor when it is fired. It does not have to be attached to the floor, but allow students to hold it down with their hands to stabilize it.

6. The Ping Pong balls will be all over the place, so I suggest you hold your competitions in a gymnasium.

7. Here are two categories of competition:

   a. distance
   Use tape to mark where the flinger should be placed on the floor. Allow each team to fire its flinger and use a tape measure to mark how far each Ping Pong ball travels. Because Ping Pong balls bounce and roll like crazy, I suggest you measure from where the ball first contacts the floor. Give each team as many flings as time allows and average the distances each team achieves. The team with the highest average distance wins the competition.

   b. accuracy
   Construct a football-style goalpost out of a large cardboard box. Tape it to the floor a reasonable distance from where the flingers will be placed. (Holding a distance competition first will help you determine what a reasonable distance should be.) Challenge each group to fling their Ping Pong ball between the uprights of the goalpost. Give each team as many flings as time allows and award three points for a fling in which the ball goes through the uprights. The team with the most points at the end wins.

Safety Tips

1. Remind students to be careful with their fingers around the mousetrap.

2. I suggest you make each team wear eye protection while they are firing their flinger. You don't want a student to catch a wayward Ping Pong ball or a piece of distingetrating flinger in the eye.

Awards and Recognition

1. Display each student's or team's flinger where they can be appreciated by other students, staff, and parents.

2. Display a chart showing the distance and/or times achieved by each flinger.

3. Give the winner(s) chocolate balls.

15 | Keep That Cube!

Challenge

Design and build a device that will prevent an ice cube from melting for as long as possible.

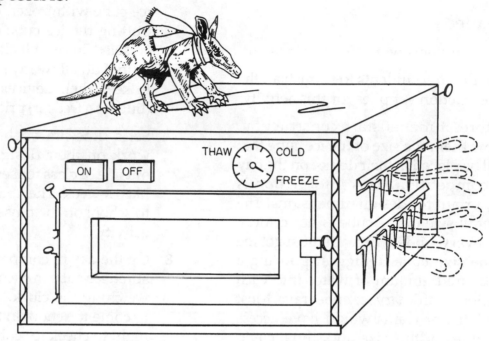

Rules

1. You must supply all the construction supplies necessary to build your device. For the actual competition, the teacher will supply a measured ice cube.

2. Your entire device can be no larger than one cubic foot.

3. To allow us to see your ice cube, there must be a "window" at least one-inch square built into your device.

4. You cannot surround your cube with ice or any material that is frozen solid.

5. You cannot put your cube inside any commercial refrigerating or insulating system or materials.

6. Your device must be built within the time limit set by the teacher.

7. The device that prevents the ice cube from melting the longest will be declared the winning design.

15 For the Teacher

The Challenge

In this activity, students are challenged to develop insulating devices from common materials.

The Rules

1. Have students work alone or in teams.

2. I require my students to furnish all the construction supplies for this activity.

3. Before students start construction, show them the size of the ice cube that will be placed in their device on the day of competition. This information is important for design dimensions. The overall size of the cube is not critical, but every student or team must get the same size cube. I suggest pouring a measured amount of water into each section of ordinary ice cube trays. Most students or teams would have access to cubes of this size and, thus, could test their systems before the competition.

4. Devices must be constructed with some type of viewing window so you can see the ice cube. An acceptable viewing window could consist of cellophane or plastic covering a hole in the top or side of the device.

5. Do not allow the use of commercial insulating materials such as fiberglass or spray foam or commercial insulating systems such as ice chests.

6. Do not allow designs that surround the ice cube with frozen material, such as putting the ice cube down in a space chiseled into a block of ice or frozen mud. I do allow my students or teams to keep their devices in a refrigerator (but not a freezer) prior to competition.

7. Set a time limit. I have my students work on their designs on their own outside of class for several days to perhaps a week. They bring their systems to school on the day scheduled for the activity.

8. On the day of competition, put a measured ice cube in each system and start watching the clock. Unless you want to come to school in the middle of the night to check ice cubes, I suggest you begin this activity as soon as possible in the morning. I have had some devices that insulated so well that a small piece of the ice cube was still intact after 14 hours.

Safety Tips

Use common sense.

16 Flame Out

Challenge

Design and build a device that uses a mousetrap to put out a candle flame.

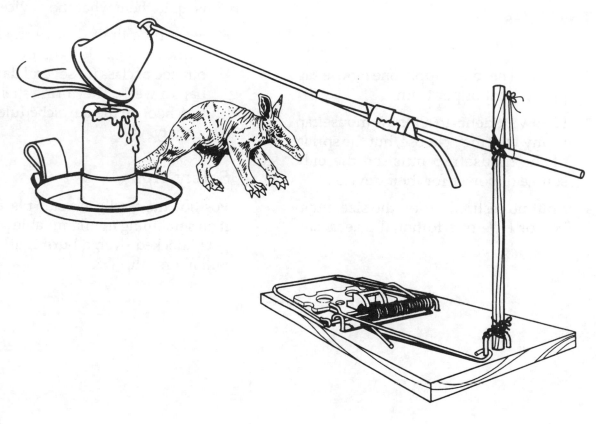

Rules

1. The teacher will supply you with one mousetrap. You must provide all other necessary construction supplies.

2. You may alter the mousetrap in any way you wish.

3. Other objects may be attached to the mousetrap or the mousetrap may be attached to other objects. There is no limit to the size or number of objects you may use.

4. The only source of energy for your device must be the spring of the mousetrap.

5. The device may be no closer to the candle than 24 inches and no farther from the candle than 36 inches.

6. The device may be placed no higher than 12 inches off the floor.

7. Your device must be built within the time limit set by the teacher.

16 For the Teacher

The Challenge

In this activity, students are challenged to turn a mousetrap into a device to snuff out a candle.

The Rules

1. Have students work alone or in teams.

2. You will need to supply one mousetrap per student or per team.

3. I allow students to alter the mousetrap in any way they choose, but the spring of the mousetrap must be the only source of power for their device.

4. I put no restrictions on the size, number, or type of additional objects stu-dents incorporate into their device in conjunction with their mousetrap.

5. I do put restrictions on how close their device can be to the candle and how high the device can be off the floor. It is up to them what the device rests on.

6. Set a time limit. I have my students work on their designs on their own outside of class for several days to per-haps a week. They bring their devices to school on the day scheduled for the activity.

Safety Tips

Position the candle in a safe place and put it on something nonflammable so that if it gets knocked over, a burn mark or a fire will not result.

17 Puncture and Pop

Challenge

Design and build a device that will use a mousetrap to activate a chain of three events that will pop a balloon.

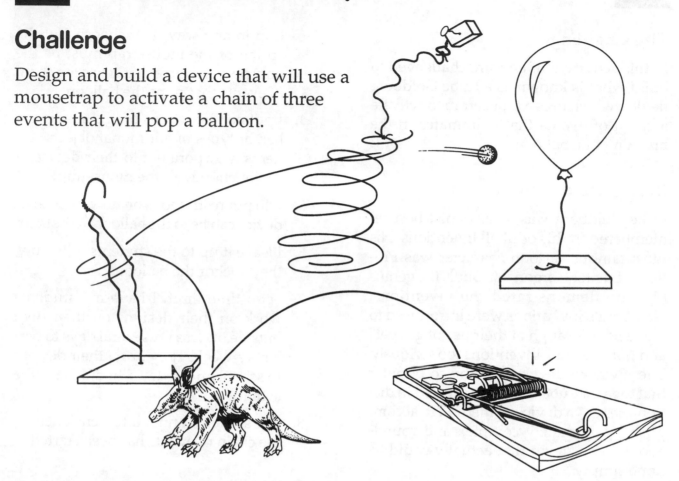

Rules

1. The teacher will supply you with one mousetrap. You must supply all other necessary construction supplies.

2. The only power source used to begin the chain of events must be the spring of the mousetrap.

3. An event would be considered as one complete unit. In other words, three dominos falling is one event, not three events.

4. Neither the mousetrap nor the balloon are part of the three events that must occur. You cannot intervene once the chain of events has started.

5. The mousetrap may be altered in any way you wish. Your mousetrap may be attached to other objects or other objects may be attached to your mousetrap. There are no restrictions to size and types of objects you can use in conjunction with your mousetrap.

6. Your device must start a minimum distance of 36 inches from the balloon.

7. Safety is an important consideration. Explain your design to the teacher before you begin construction. Don't be disqualified by a dangerous design.

8. Your device must be built within the time limit set by the teacher.

17   For the Teacher

The Challenge

In this activity, students are challenged to build what is known as a Rube Goldberg device, which uses a mousetrap to activate a chain of events that culminates in the breaking of a balloon.

Background Information

Rube Goldberg was a cartoonist best remembered for his oddball inventions. His most famous cartoon character was Professor Lucifer Gorgonzola Butts, the genius (?) who demonstrated the inventions. Goldberg's inventions were introduced to make people laugh at their use of gadgets and not take new inventions too seriously. These wacky devices became so popular that today a Rube Goldberg is listed in dictionaries as "a device or method to accomplish by extremely complex and roundabout means a job that actually could be done simply."

The Rules

1. Have students work alone or in teams.

2. You will need to supply one mousetrap per student or per team.

3. I allow students to alter their mouse-

trap in any way they choose, but the spring of the mousetrap must be the only power source that starts the chain of events that pops the balloon.

4. I put no restrictions on the size, number, or types of additional objects students incorporate into their device in conjunction with the mousetrap.

5. I do put restrictions on how close their device can be to the balloon at the start.

6. I leave it up to the students as to how they anchor the balloon.

7. Set a time limit. I have my students work on their designs on their own outside of class for several days to perhaps a week. They bring their devices to school on the day scheduled for the activity.

8. The student cannot intervene once the chain of events has been started.

Safety Tips

Most, if not all, of the designs will use sharp objects such as nails or pins to pop the balloon. I strongly urge you to evaluate each design from a safety standpoint before you let students even begin construction. Discourage designs that would use flying projectiles with sharp objects.

18 Hole in One

Challenge

Design and build a device to propel a golf ball into a paper cup at least 36 inches away. While traveling, the golf ball must put out a candle and set off a mousetrap.

Rules

1. The teacher will supply the following:
    - 1 mousetrap
    - 1 golf ball
    - 1 candle
    - 1 paper cup
    - 2 rubber bands

    You must supply all other necessary construction supplies.

2. The only power source used to propel the golf ball must be the rubber bands.

3. Only regulation golf balls are allowed—no alterations.

4. There are no restrictions on the location of the candle, mousetrap, or paper cup.

5. You cannot intervene once the golf ball has started moving.

6. Your device must be built within the time limit set by the teacher.

18 For the Teacher

The Challenge

This is another Rube Goldberg activity. Students are challenged not only to propel a golf ball into a cup but to put out a candle and set off a mousetrap at the same time.

The Rules

1. Have students work alone or in teams.

2. You will need to supply the following materials per student or per team:

    • 1 mousetrap

    • 1 golf ball
    (Local golf courses or parents who golf may donate their old bruised and battered golf balls.)

    • 1 candle
    (I prefer small birthday candles because they are cheap and easy to obtain.)

    • 1 paper cup
    (I use waxed paper cups, generously donated by the school concession stand.)

    • 2 rubber bands
    (The exact size is not critical, but all students or teams should get the same size.)

3. The only restriction is that the rubber bands must be the only power source used to start the golf ball rolling. I place no restrictions on the size, type, or number of objects used in the design or on the location or position of the candle, mousetrap, or cup. However, the student cannot intervene once the golf ball starts moving.

4. Set a time limit. I have my students work on their designs on their own outside of class for several days to perhaps a week. They bring their designs to school on the day scheduled for the activity.

Safety Tips

Discourage designs that have sharp, pointed projections, and avoid those that employ flying projectiles. If you feel there is any question about safety, require students to have their designs evaluated by you before construction begins.